SCIENTIFICALLY THINKING

SCIENTIFICALLY
THINKING

How to Liberate Your Mind,

Solve the World's Problems, and

Embrace the Beauty of Science

STANLEY A. RICE

 Prometheus Books

59 John Glenn Drive
Amherst, New York 14228

Published 2018 by Prometheus Books

Cover image of silhouette of woman © Kazitafahnizeer/Shutterstock
Cover image of science © Zoezoe33/Shutterstock
Cover design by Liz Mills
Cover design © Prometheus Books
Cartoons by Leslie Gregersen

Inquiries should be addressed to
Prometheus Books
59 John Glenn Drive
Amherst, New York 14228
VOICE: 716–691–0133 • FAX: 716–691–0137
WWW.PROMETHEUSBOOKS.COM

22 21 20 19 18   5 4 3 2 1

Library of Congress Cataloging-in-Publication Data

Names: Rice, Stanley A., 1957- author. | Gregersen, Leslie.
Title: Scientifically thinking : how to liberate your mind, solve the world's problems, and embrace the beauty of science / Stanley A. Rice ; cartoons by Leslie Gregersen.
Description: Amherst, New York : Prometheus Books, 2018. | Includes index.
Identifiers: LCCN 2018034609 (print) | LCCN 2018038614 (ebook) | ISBN 9781633884717 (ebook) | ISBN 9781633884700 (hardcover)
Subjects: LCSH: Science—Popular works.
Classification: LCC Q162 (ebook) | LCC Q162 .R4945 2018 (print) | DDC 500—dc23
LC record available at https://lccn.loc.gov/2018034609

Printed in the United States of America

CONTENTS

SECTION I: THE ADVENTURE OF SCIENCE

SECTION II: LEGACY OF AN APE'S BRAIN

SECTION III: BIG IDEAS

SECTION IV: THE ROLE OF SCIENCE IN THE WORLD

LIST OF ILLUSTRATIONS

AUTHOR'S NOTE

Many of the concepts in this book lend themselves directly and immediately to classroom learning activities, for which there is insufficient room in this book. Educators may visit the author's website, http://www.stanleyrice.com, to find learning activities based on chapters in this book, as well as a collection of photographs that are suitable for classroom discussion. Some chapters for which there was no room in this book can also be found at this site. These resources are available without charge.

ACKNOWLEDGMENTS

The author would like to thank literary agent Rita Rosenkranz and editor Steven L. Mitchell. Christopher Moretti produced figure 7-2. The author also appreciates interesting discussions of the topics in this book with graduate students in his research methods classes at Southeastern Oklahoma State University in Durant over the past decade and a half.

The author also wishes to acknowledge the many fellow scientists and science educators who have helped him develop the skill of scientific thinking during the last forty years. Rather than to just tell him what they considered to be the truth, they helped him figure it out for himself, sometimes by letting him make his own mistakes before finally succeeding. This list would be very long, but the author particularly wants to thank R. Edward Dole, an instructor at the University of Illinois at Urbana-Champaign, who supervised the author's management of the botany laboratories while he was in graduate school.

The author also wishes to acknowledge Lee Rice and Benjamin Gennetay for updating his author website to include the resources described in the author's note.

WE NEED SCIENCE, AND WE NEED IT NOW

We humans can be justifiably proud of our brains. No other species has ever had such large brains and so much intelligence. But there is a major problem. Our brains evolved. The brains of our most successful ancestors were not necessarily the ones that understood the truth but the ones that allowed their possessors to prevail in the struggle for existence. As a result, our brains did not evolve to reason but to rationalize, or to see the truth but to create it. Our brains are the playthings of bias and illusion. We use our brains to manipulate others and to deceive ourselves.[1] We often see what we want to see rather than what is really there. All of us.

Illusion and bias are not necessarily bad things. Illusion and bias can be simple and useful. Without these illusions and biases, our brains might be overwhelmed by the complexity of the world. Illusions and biases have allowed our brains to make quick decisions in order for us to survive a sudden threat or to take advantage of a sudden opportunity. An attack by a predator or an enemy is not the best time for rational thought. Our ape brains have been immensely successful in large measure because they can make profitable use of reality and truth but are not constrained by them.

Our biased and deluded brains served us well enough in the past. We did not need to understand the truth so long as we could be successful in the game of evolution. But today, our biases and illusions put us at risk of worldwide catastrophe. We are drowning in yottabytes of information. We do not need more information but a new way of thinking that will liberate our minds from bias and illusion.

Some of the mistakes created by our biased brains affect us only individually. Mistaking correlation for causation can make us waste time and money on health fads that are illusions. But other mistakes that we make collectively can endanger the whole world. Our species is the victim of its own success, and now we have overrun the earth. Every natural habitat, even those we have set aside as parks and reserves, has been altered by our presence and our economic and political activities. Errors made by one individual, one corporation, or one nation can now affect the whole world. Today, all individuals and all countries and all economies are so interconnected that if one group acts like cavemen, they can drag much of the rest of the world down with them. They can be religious extremists trying to wage holy war, or they can be executives of "too big to fail" corporations who use misinformation to increase their profits, or they can be politicians who create their own "facts" in order to win campaigns, and their effects will be felt worldwide. With over seven and a half billion people in the world, there is no room for delusion anymore. Having an ape brain couldn't happen at a worse time. Illusions and biases served us well in the past, but today they have brought us to the brink of disaster.

One example of how our brains mislead us is the illusion of the cornucopia of nature. It just seems natural to assume that the world is big enough to supply all of our desires and absorb all of our wastes. Maybe it was like that back in the Stone Age. When a caveman threw bones away—"away" meaning outside the camp—the great maw of nature could decompose them. But there are now over seven billion of us, each of us using immense amounts of material and energy and generating toxic wastes. There is no such place as "away" where our wastes can be safely dumped.

Perhaps the most important example of this illusion is global climate change, a topic that I will revisit repeatedly in this book. You can't see global warming. You have to deduce it from evidence. All you can see is weather; you can't see climate. Climate is the long-term and broad-scale average of weather. As Mark Twain is reported to have said, climate is what you expect, weather is what you get. Climate is the weather you expect to see in an "average year." But as the old farmer said, he'd been farming for forty years and had only seen two average years.

We can get the answer to almost any question with just a few keystrokes and the swoop of a computer mouse or the swipe of the screen. You would think that having all of this information at our fingertips would make it easier to find the truth. But this is not the case, for several reasons.

One reason that more information does not lead to a better understanding of the world is that, no matter how much information we have, we can only think about one thing at a time. For a moment, it seems we can take hope in the prodigious capacity of the human brain. Our brains are more complex than any computer. In 2013, a supercomputer consisting of eighty-three thousand processors took forty minutes to mimic what a typical human brain can do in one second.[2] But each of us can still only think about one thing at a time. Herman Melville wrote in 1851, "So long as a man's eyes are open in the light, the act of seeing is involuntary; that is, he cannot then help mechanically seeing whatever objects are before him. Nevertheless, as anyone's experience will teach him, that though he can take in an undiscriminating sweep of things at one glance, it is quite impossible for him, attentively, and completely, to examine any two things . . . at one and the same instant of time . . . in order to see one of them, in such a manner as to bring your mind to bear on it, the other will be utterly excluded from your contemporary consciousness."[3] Your brain is doing lots of things subconsciously this very moment. Your brain stem is controlling your body temperature, your heartbeat rate, and your breathing. Your cerebellum is keeping you from falling out of the chair in which you are probably sitting at this moment. But your conscious mind can only pay attention to one thing at a time, no more and no less than a Cro-Magnon hunter-gatherer. You could say that, in this sense, nothing has changed much in the last twenty thousand years.

Another reason that more knowledge does not lead to greater understanding is that, as explained above, the human mind does not look for truth in data but rationalizes it—no matter how many or how few data there are. The easiest thing for our brains to do, when overloaded with information, is to pay attention only to the information that agrees with what we already believe. We have, at our fingertips, troves of information that can *free us* from our ancient biases, but instead we use this information selectively to *reinforce* our biases. Perversely, we become more mistaken when we have access to more information.

And not only do we use our brains to deceive others, but we also use our brains to deceive *ourselves*. Our brains take the data of reality and alter our perceptions to create self-deception, as explained by Robert Trivers.[4] On the bad side, by actually believing the false things that we claim to be true, we become more convincing liars. On the not-so-bad side, by actually believing false things, we can delude ourselves into being happy despite unalterably dismal circumstances. Either way, believing false things can sometimes make us more successful in the evolutionary struggle. And we have to really believe them, not just pretend to. Our brains did not evolve just to rationalize but also to believe that rationalization is reason. Our brains are the amphitheaters of delusion.

We do not need more information about the world. What we need desperately is some way to escape the biases and illusions created by our evolved brains. We need some good news, and we need it now.

But there is good news. This way of thinking already exists. It is the scientific method. Although science has existed in its modern form for only a few centuries, and although most people in the world do not avail themselves of it, science has led to immense breakthroughs in understanding the world. In just a few centuries, we have come to understand the universe, the earth, and ourselves as we never could, and never could imagine, before.

To save ourselves and the world, do we all have to become scientists? This cannot happen, and if it had to, there would be no hope for any of us. Fortunately—and this news is good enough that you can put it on a placard on your wall—you don't have to be a scientist to use the scientific method. The scientific method is based on mental processes that come naturally to all of us. Most people realize that using "the process of elimination" is just "common sense." We have an instinct for science, just as we also have biases and illusions. The scientific way of thinking is lurking in our minds, if we can just give it a chance to emerge. The first purpose of this book, then, is to convince you that we desperately need to think scientifically—not just scientists but also all of us.

Some people have the impression that scientists have discovered the secrets to the universe. But this belief carries two insidious unspoken messages. The first is that scientists are a species apart, a brilliant breed of super-hominids whose thought processes a non-scientist can never hope to understand. The second is that scientists want everybody else to think they are smart so that nobody will question what they say, which leads directly to the antipathy that many citizens feel toward academics in general and scientists in particular, that scientists have, for example, hoodwinked everyone else about evolution and global warming. The result is a widespread disregard for scientific evidence.[5] But most scientists invite everyone who wishes to apply her innate critical skills to see the evidence and draw conclusions for herself.

The second purpose of the book is to show you that the scientific way of thinking is not only instinctual but is also surprisingly simple. It is something that anybody can do. Scientific thinking is basically just organized common sense. The hard part is not the science itself but keeping our biases and illusions out of it. I am a scientist and a science educator, and I want to help you get started on the journey of science.

Here is the way we can think scientifically. First, we create a statement about cause and effect, a claim about how something works. We call such a statement a *hypothesis*. Second, we gather evidence that *tests* the hypothesis. The evidence must be something that is accessible to anyone (external), not just to the person who makes the claim (internal). This evidence will allow us to determine that the hypothesis is true, that it is false, or that we do not yet know. Third. . .

There is no third. That's it. That's science. We test claims by means of evidence. The hypothesis, if it withstands the test, makes sense of an otherwise confused mass of facts. One hypothesis can lead to another, with the eventual result that a whole framework of hypotheses—now called a *theory*—helps make sense not just of a set of facts but also of the whole world.

But suppose we reach the wrong conclusion. No problem. Since the evidence is external, someone else can come along and present better evidence, which may lead to a different conclusion.[6] In contrast, faith is an internal experience. Many people cling to religious beliefs in the absence of, or even in opposition to, external evidence. But science is external. Science is public. What one person claims, another can test.

Yes, the scientific method is liberating, all right, but it liberates your mind by constraining it. Your mind is free to float untrammeled through all of the clouds and rapids and cliffs of bias and illusion. In order to make your mind seek the truth, you have to keep it from running off in other directions. Think of an ox dragging a cart through the mud. The ox could go in any direction, leaving the cart in the mud. But the yoke constrains the movement of the ox. Science is that yoke: it allows the ox (your mind) to pull the cart of knowledge forward through the mud of confusion.

We scientists have brains that are just as biased and as plagued with illusions as anyone else's. But we have developed a system that allows us to escape from them, most of the time. Most of this book is about the kinds of errors our ape brains can lead us into, and what the scientific method does about them. You cannot even measure something without bias. For example, how do you know that the measure is valid? Is the gross domestic product a valid measure of national wealth? Even things as simple as accurately measuring the height of a person or how hot the weather is outside can be a challenge. Not only do bias and illusion make it difficult for you to reach the correct conclusion; they also even contort the very process of measurement itself. You cannot even trust your senses when making measurements!

I not only explain these challenges, but I also give examples of major studies in which scientists have used very, very creative methods of avoiding bias and illusion. When you read about these studies, you will forever reject the notion that science is a plodding ramble through dusty piles of facts. Scientific creativity is exceeded only by the prodigious and stupefying creativity of the natural world itself. Scientists think big because their subject is big.

There is more good news. Not only can any normal human being participate in scientific thinking, but also every normal human being begins childhood with that ability. The tiniest infant looks around and observes everything, first the nearby things such as his mother on which he can focus his eyes, and, later, things that are farther away. Although an infant is also subject to inchoate forms of bias and illusion, the infant spends most of his time applying the scientific method to understanding the world. What happens if I push the plate to the edge of the tray? How do I tell which foods are yummy and which are yucky? What happens if I run into a wall? What happens if I scream? And there is more. Children are curious about not just the room and the toys but also about what we call the natural world—trees and birds and rocks. It is only later that an older child may develop the attitude that the natural world is boring.

The ones who never develop this attitude often grow up to be scientists or science teachers. The biggest challenge in science education is to keep kids from ever becoming bored in the first place. The National Research Council may have exaggerated just a little in the title of their report, *Every Child a Scientist*,[7] but not much. Science is a creative adventure, not entirely different from other kinds of human thought, such as creative writing.

You can experience some of the joy by reading science books and also by going on a field trip led by an enthusiastic professional or amateur scientist. You can even become an amateur scientist yourself by joining one of the nationwide or worldwide networks of citizen-scientists who gather data and submit them to large online databases. The data can be anything from bird sightings to human aging to light pollution to human sexuality. As much fun as science may be to read about, it is much more fun to do, and you don't need to be a credentialed scientist to contribute to this worldwide scientific quest for understanding. You can be an amateur scientist. People often use the word *amateur* disparagingly— but it comes from the Latin word for *love*. Amateurs are people who do something because they love it, not because it is their job.

There is even more good news. Not only can the scientific method lead our minds, appropriately constrained, to find the truth, and not only is the scientific method based on an instinctual process, but science can also lead us into a world of beauty. It does so in several ways. Consider these examples.

- First, there is a deep sense of beauty in knowing that the world makes sense. The world is not what Carl Sagan called, referring to the nonscientific view of the universe, a "demon-haunted world" in which anything can happen any time based on the whim of a deity.[8]
- Second, there is beauty in knowing that there is something we might be able to do about our problems. Before we knew that germs caused many diseases, there was not much we could do about them. But now we have, within a couple of years of its first emergence, identified the Zika virus and we know what it does and how it does it. It is not a pretty picture, but our knowledge allows us at least a chance of slowing down its spread and dealing with its consequences.

- And finally, science allows us to begin recognizing the diversity around us. To someone who has never studied trees, a tree is a tree is a tree. But once you begin to study trees, the world becomes so much more interesting because you notice how many kinds there are and how each one of them is a little different in the way it functions in the natural world. Cottonwoods grow fast and die young; oaks grow slowly and die old: once you realize this, a forest is no longer a backdrop but is, in itself, an interesting story.

Our brains evolved to rationalize, but the ability to reason is one of the components of our ape-brain process of rationalization. And when you grasp this simple truth, you are ready to begin thinking scientifically. Nor do you have to believe everything every scientist says as if it is handed down by God, any more than you have to believe everything a politician or a preacher says. You can apply the scientific method to everything and draw your own conclusions. This is perhaps the most exhilarating liberation of which the human mind is capable.

The scientific method is organized common sense. This sounds simple enough. But then it gets complicated. In order to test hypotheses, scientists sometimes have to do absurdly beautiful experiments that make you laugh and then make you think. And it is difficult to deal with bias and illusion. Scientists have to go through a lot of effort, and spend a lot of money, to design experiments that avoid bias and illusion, not only on the part of the experimenter but of the experimental subject as well.

- Bias and illusion influence even the very process of measurement itself: how accurately do we measure something, and how big of a sample do we need?
- Our brains tend to see straight lines while nature throws us curves. This is particularly important with the nonlinear economies of scale and threshold values. Even the explosion of evolution and the death of Mars were nonlinear processes.
- Our brains see everything as categories (often just two) while in nature most of what we see is continuous variation.
- Our brains see everything as simple cause and effect while in nature an effect can also be a cause, and an effect can have multiple causes.

- Our minds have the bias of agency: that is, when something happens, an intelligence must have deliberately caused it to happen. Because we are intelligent, we see intelligence every-where. How can you tell whether an animal is intelligent or not?
- You might be perfectly capable of making measurements, but if the measurements do not represent a fair sample of the diversity of reality, or are in fact measuring something other than what you think they are, they are invalid and misleading.
- Our brains have a confirmation bias in which we literally see what we expect to see rather than what's really out there.

Science can liberate us from the errors to which even the best human brain is vulnerable. But it has also generated some important Big Ideas that have transformed our whole view of the universe and ourselves. Science is very different from religion, which appears to be a human instinct. Finally, knowing what we know about the world (about everything from global warming to the dangers of tobacco), we scientists cannot sit silently and let people who have it all wrong mess up this beautiful and, as far as we know, unique world. And this leads us on a beautiful venture.

Join me in exploring how the process of science liberates us from illusion and bias. Come laugh with me at the creative ways in which scientists find creative ways to test their hypotheses. And come and see that the scientific way of thinking is not completely different from the common sense and the creativity that we have always known and loved.

DO IT YOURSELF!

Many chapters in this book will have a "Do It Yourself!" box that will suggest scientific activities that you can do.

SECTION I

THE ADVENTURE OF SCIENCE

Science is an adventure of discovery, not a pile of facts. It is an adventure of discovery not too different from other ways in which humans explore ideas and, if they are lucky, figure things out. Pseudoscience can masquerade as science. However, its purpose is not discovery but propaganda, not liberation of the mind but its conquest.

CHAPTER 1

SCIENCE AND HOW TO RECOGNIZE IT

Science is a way of knowing. The human species has developed many ways of answering questions involving religion, tradition, deductive philosophy, etc. Science is a relatively new way of finding things out, taking its modern form in the Renaissance.[1] Its roots can be traced back even further, however. The eleventh-century French priest Peter Abélard defended the right to doubt the "truths" that we have inherited from the past. He said, "It is by doubting that we come to investigate, and by investigating that we recognize the truth." (He got in trouble for it.) Science has proven to be a very powerful, very effective, and very enjoyable way of finding things out, as explained by physicist Richard Feynman and biologist Richard Dawkins.[2]

As noted in the introduction, a *hypothesis* is a statement about cause and effect, a story about the way things work. Science does not test just any hypotheses. Scientific hypotheses must have the following characteristics:

- *Physical.* Science studies only things that happen by the laws of nature. Science does not study miracles. This does not mean that miracles never happen. Some scientists privately believe in miracles but will never include them in their scientific work.
- *Repeatable.* Singularities (processes that happen only once, of which miracles would be an example) cannot be studied by science. Science can, however, investigate a nonrecurring event if that event has left behind enough evidence. Scientists can study

the asteroid impact that occurred sixty-five million years ago and caused the end of the dinosaur era. No such massive asteroid impact has occurred since that time, but it was an entirely natural process and *could* happen again.

- *Measurable.* There must be some way of not just detecting the object of study but, if possible, of also quantifying it. Scientists really do try to reduce everything to numbers but only for analysis. Scientists revert to real, human concepts when interpreting and applying the results.

- *Falsifiable.* Any scientific statement must be capable of disproof. If there is no possible set of observations that could ever prove a hypothesis wrong, then it is not scientific.

The first three points impose the limits of reality on science. Science may tell us that things we want to believe are physically impossible. Believing in the impossible is for religion, not science.

The fourth point is particularly important. Many people and organizations who claim to have scientific credibility champion hypotheses that are unfalsifiable. That is, no matter what data may come along, they find some way to incorporate them into their system of beliefs without changing those beliefs. Religion has done this for a long time. No matter what happens in the world, religious people find some way of explaining it away as being God's will. But scientists are not supposed to do this.

Perhaps the most famous philosopher of science, Karl Popper,[3] realized that Einstein's theories of relativity were falsifiable, while Sigmund Freud's theories of psychoanalysis were not. No matter what he observed, from cigars to happiness, Freud found some way to "explain" it by psychoanalysis. It was not possible to specify anything that might come forth from the human brain that would make Freud say, "Well, I guess I was wrong." But Einstein's theories could be put to the test. In 1919, Sir Arthur Eddington did just that when he demonstrated, during a total solar eclipse, that the gravity of the sun bent the light from stars. Indeed, such "gravitational lensing" of starlight is still used to determine the mass of faraway stars.[4] Had this bending not occurred, Einstein's theories, at least in the form that then existed, would have been discredited. It could have been back to the drawing board for Einstein but never for Freud.

The same disregard for falsifiability seems to underlie the assertions of those few scientists who reject the consensus on global warming. Most climate scientists, and scientists in general, accept the hypothesis that human activities (which include the release of carbon dioxide and methane gas into the air, and the destruction of forests and prairies, as well as the indirect effects of these activities) have caused the earth to become warmer during the past 150 years, more than natural processes, such as variation in the intensity of sunlight or the dust and gas from volcanoes, could have produced during that time. In 2017, the Government Accountability Office estimated that, during the preceding decade, extreme weather and fire events had cost the federal government $350 billion and that global warming will make the problem worse in the immediate future. Note that this is just the cost incurred by the federal government, not by state and local governments, by corporations, or by individuals.[5] Credit agencies, such as Moody's Investor's Service, have begun to take preparedness for global climate change into account when assessing the creditworthiness of US municipalities; if a city has no plans to deal with global warming, it may receive a lower credit rating.[6] The alternative hypothesis would indicate that if global warming has occurred, it has been largely due to natural processes, not human activity. The US Global Change Research Program, a joint project of thirteen federal agencies mandated by the Global Change Research Act of 1990, has issued a series of detailed reports in which they calculate how much global warming has occurred, how much human processes have contributed to that warming, and how much global warming would have occurred in the absence of human processes. The scientists on this panel concluded that, indeed, natural processes contributed to global warming but that human processes contributed well over ten times as much.[7]

But no amount of evidence seems to satisfy the "denialists" or "climate deniers." To them, it appears, as long as any doubt about any of the measurements remains possible (and there is always doubt about any measurement), then we must conclude that humans are not causing global warming. No matter what evidence climate scientists may present, it will never be good enough to satisfy the deniers. Climate change deniers *could* be scientific, but, because their hypotheses are unfalsifiable, they are not.

WHERE SCIENTIFIC HYPOTHESES COME FROM

The first step in scientific investigation is to generate a hypothesis. Here is an example. I am a botanist. One time I designed a class project that tested a hypothesis about the way plants grow. I hypothesized that plant roots would proliferate in layers of high-nutrient soil more than they would in layers of nutrient-poor soil. You could think of my hypothesis as a story: when roots encounter rich soil, they branch out and form a network that draws in the nutrients, but when they encounter poor soil, they grow straight down through it, increasing their chances of finding rich soil farther down.

A brain generates hypotheses about things that it knows the most about. The more you know about something, the more likely you are to generate a hypothesis that is worth investigating. How can you generate a hypothesis about the growth of roots if you have never dug underground to look at them?

But the act of generating a hypothesis need not be entirely conscious. Sometimes hypotheses are just hunches that spring out of the subconscious mind. Our subconscious minds are doing calculations and estimates, and they present an executive summary of their findings to our conscious minds, which may be unaware that any calculations or estimates may have been done. Our subconscious minds form and test hypotheses, though in a manner less disciplined than that employed by our conscious minds. That is, our subconscious minds are not merely the haunt of primitive animal emotions. They often practice a primordial form of scientific inquiry, of which our conscious minds see only the tip of the iceberg.

One example of this is a pharmaceutically active plant extract that I discovered. It seemed to be merely a chance discovery. I made an extract from some twigs of a rare species of tree that I study, and discovered that the extract kills staph bacteria. I had no reason at all to believe that it would; in fact, when I used the extract against other kinds of bacteria, it did not work. But it was not entirely a guess. I knew that many pharmaceutical compounds originally came from plants and that some of them were "chance" discoveries.[8] Therefore, my discovery was not as random as it seemed.

Sometimes, but only rarely, hypotheses emerge from dreams. The chemist August Kekulé literally dreamed up the structure of the benzene molecule. He dreamed of a snake forming itself into a ring by biting its own tail. This dream would have been of no use to him if he had not been trained in chemistry. But to him the dream suggested the hypothesis that the core of a benzene molecule was a ring of carbon atoms. Whether in hunches or in dreams, the subconscious mind can suggest hypotheses.

When generating hypotheses, scientists attempt to take every possibility into consideration. This is, of course, impossible. Nobody, scientists included, can ever be sure whether all possibilities have been considered. That is where Sherlock Holmes got it wrong. He said that when all other competing hypotheses have been eliminated, the one remaining hypothesis, no matter how incredible it may sound, must be true. Modern scientists know better than to believe this because, as it turns out, there are almost always alternate hypotheses out there in the universe that we have not thought of yet. There are always "unknown unknowns." Neither a scientist nor anyone else can say, "I have examined all the possibilities." Perhaps this is why the science of evolution did not emerge until a couple of centuries ago. No one could have thought of evolution as even a possibility until scientists had learned about genetics and about the populations within which natural selection occurs. Only then did the hypothesis of evolution become thinkable and testable.

SIMPLIFIED MODELS OF REALITY

The world is too complex for us to pay attention to everything that happens; we have to use hypotheses that are *simplified* models of reality. Scientific research is carefully crafted to isolate just the process that the scientist is studying and to eliminate everything else. This is not easy to do; as a matter of fact, it is perhaps the hardest part of scientific research.

Let me give a simple example of this. There is a species of wildflower, which lives only one year and reproduces only from seeds, and which is very rare out in the dry oak woodlands where I live. But the soil is full of its seeds. Along comes a big fire, which burns down the whole forest. The following spring, the wildflower seeds sprout and the flowers bloom profusely, covering whole acres with astonishing purple flowers. There is obviously something about the fire that causes the seeds to germinate. Now, the woodland is a very complex place. And fires are very complex. Seeds that sprout after fires are exposed to heat and smoke during the fire, and in the season after the fire the seeds experience brighter light (because the larger plants have been burned away) and a flush of nutrients from the ashes. To incorporate all of the complexity of the fire into an experiment will contribute nothing to understanding what caused the seeds to germinate. Instead, my research assistant and I needed a simplified experiment. Our hypothesis was that chemicals in the smoke caused the seeds to germinate.

PUTTING IT TO THE TEST

Once a scientist has a hypothesis, she must put it to the test. This idea, of testing hypotheses rather than just accepting received wisdom, is a hallmark of science that goes back at least as far as Francis Bacon's *Novum Organum*.[9]

To test a hypothesis, you need to have something to compare it to. That is, you have to know what your measurements would be like if your hypothesis is not true. Your *null hypothesis* states what you expect to see if your main hypothesis is wrong. (*Null* means *nothing*.)

Sometimes scientists test hypotheses by conducting an *experiment*. In an experiment, the scientist is actually in control of what is going on, which helps to eliminate as many extraneous factors as possible. The *treatment* is based on the hypothesis, and the *control* is based on the null hypothesis.

To test the root-growth hypothesis mentioned previously,[10] I had my students grow some sunflower seedlings in glass cylinders with layers of rich and poor soil. We watched (and measured) the growth that was visible from the outside as the roots penetrated through the layers. Once the roots had grown to the bottom of each cylinder, we were able to compare the amount of root growth in the rich soil to that in the poor soil. The layers of good soil were the treatments, and the layers of poor soil were the controls. We also used cylinders of just good soil, or just poor soil, to which we could compare our results. We found that roots grew in the good soil more than in the poor soil (figure 1-1).

Figure 1-1. A simple experiment that demonstrates that plant roots proliferate in rich soil (here, dark loam potting soil) more than in soil that is poor in nutrients (here, white perlite).

Students could see and measure root growth through the glass. In some of the glass columns, the seedlings first encountered a thick layer of loam, while in others they first encountered a thick layer of perlite, thus correcting for the "sequence effect" (see chapter 12), in which the first encounter is the decisive one. Even when perlite was the top layer, it was necessary to use a *thin* layer of loam. Loam contains microbes that inhibit the growth of fungi that would otherwise kill seedlings growing on pure perlite. The controls were pure loam or pure perlite. This experiment used beans, whose roots did not distinguish between loam and perlite. When the experiment was repeated with sunflowers, the roots grew more in the loam than in the perlite.

To test the smoke-and-seed-germination hypothesis, we placed seeds on wet paper in little clear dishes under a light.[11] There were about twenty dishes. Half of the dishes just had plain water. The other half had water that had been infused with smoke. The dishes with plain water were the control, and the dishes with smoky water were the treatment. We found that the seeds in plain water did not germinate. That is, light and water were not enough to make them germinate. But many of the seeds in the smoky water germinated, indicating that the smoke caused germination. (We could have used just one plate for each treatment, but scientists always replicate their experiments because you never know what might happen to just one dish—a fungus might start growing in it, or the lid might get knocked off, who knows? And we had about twenty-five seeds in each dish.) These plates were a very simplified subset of the complex reality of a forest fire.

The only difficult part of our experiment was trying to get smoke infused into water. Some scientists use specialized apparatus to do this, such as pumps, test tubes, and glass pipes. But my assistant and I used a hookah pipe. We burned oak wood in the bowl and drew the smoke through the water reservoir using a nasal aspirator. We did this rather than puffing on the smoke because it took three hours, and who wants to smoke oak wood for three hours? Most people who use such pipes for tobacco and other things want the smoke, but we wanted the water. After three hours, the water was a foggy amber color. Our experiment cost nothing, besides the dishes we already had sitting around in the lab, because my assistant already had a hookah pipe. I never asked her what she normally used it for, but I will simply note that her relatives live out in the woods of rural eastern Oklahoma.

If it turns out the hypothesis is right, that's wonderful, and the scientists can go write a paper about it. (We did.) But if it turns out that the hypothesis is wrong (that the null hypothesis is right), the scientist has to revise the hypothesis and try again.

Science is, in a way, a process of elimination. You test one hypothesis after another until you find one that works. It is not just scientists who do this. Let me give you an example. Did you ever listen to *Car Talk*? You know, the radio show where people call in with questions about their cars? The hosts, Tom and Ray Magliozzi, were pretty funny. But they also used the scientific method all the time. They would probably have laughed if someone called them scientists, but they were. So let me use an automotive example to illustrate the process of science.

Suppose the oil light comes on in your car. You might lose your cool for a few moments, but pretty soon your brain (yes, yours) switches to a scientific mode of thought. Your first idea (hypothesis) might be that your car does not have enough oil. Then you test this hypothesis. You check the oil level with the dipstick. If the oil level is low, you get more oil. But suppose the oil level is not low. Your next idea (hypothesis) may be that the oil pump is defective. So you have a mechanic check your oil pump. If that is the problem, you just get a new oil pump. But suppose the oil pump is working. Perhaps your next idea (hypothesis) is that the oil filter is clogged. So you have your mechanic check the filter. And if that isn't the problem, then perhaps the oil sensor itself is malfunctioning. You eliminate one possibility after another (figure 1-2).

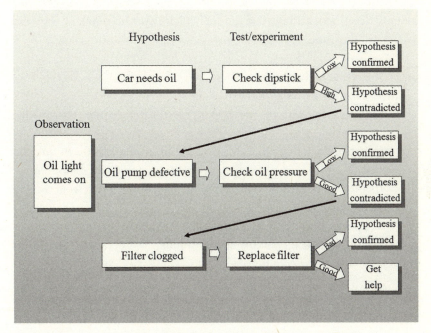

Figure 1-2. Are auto mechanics scientists? Here is a flow chart of sequential tests of hypotheses about why the oil light comes on in your car, as explained in the text.

You have to make sure that you do your tests or experiments properly. For example, you need to turn the engine off and wipe the dipstick, then reinsert it.

What I have just described is the scientific method. You have an observation that you need to explain. You propose a hypothesis, and then you test it. You may end up testing a series of hypotheses. Eventually, unless the problem becomes so complex that you decide it is not worth the time and money to solve it, you will reach an answer. It is common sense. The process of elimination. Science.

As I noted in the introduction, there is no single list of steps that constitute the Scientific Method. I said that there were two steps: propose a hypothesis, then test it. But we can expand these two steps. Here is my slightly longer list of the basics, the first three of which this chapter has already described, and the last four of which make up much of the next section of the book:

- We specify a hypothesis.
- We specify a null hypothesis also.
- We design a test of the hypothesis that eliminates extraneous factors.
- We spend a lot of time thinking about all the ways we could be wrong, so that we can prevent these things from happening *before* the research rather than trying to figure out what went wrong afterward.
- We perform the research and gather the data.
- We do statistical analyses on our results.
- Then we draw conclusions.

Scientists need hypotheses not just to reach a conclusion but also to do any work at all. Without hypotheses to guide us, we scientists would run around in the world of exciting and fascinating data like a puppy in a room full of new butts.

Sometimes it is difficult to decide which of two competing hypotheses is the null hypothesis. Consider, for example, the controversy over the use of herbicides in agriculture, particularly glyphosate, the active chemical in a Monsanto herbicide. You didn't know there was a controversy? This is because, in the United States, there isn't much of one. The hypothesis is that glyphosate is, beyond any doubt, unsafe (that is, it has adverse effects on human or animal health); otherwise we should accept the null hypothesis that it is *safe*. The burden of proof, therefore, is on those who claim that it is unsafe. But, to many scientists and citizens in Europe, the hypotheses are reversed. The hypothesis is that Monsanto must prove that glyphosate is safe beyond any doubt; otherwise we should accept the null hypothesis that it is *unsafe*. The burden of proof then falls on Monsanto.[12]

In all of the foregoing, I have assumed that null results ("this didn't work") are bad. But actually, as Stuart Firestein explains in *Failure: Why Science Is So Successful*, the null results are important.[13] Scientists seldom report experiments in which the hypothesis fails. But scientists should, Firestein claims, because other scientists should know about which hypotheses have already failed, so as to not repeat them, or perhaps to improve on the techniques of earlier scientists. Firestein speculates about a possible online clearinghouse for null results that scientists can consult. Of course, null results do not make as compelling a narrative as results that confirm a hypothesis. To read one of them would be like reading a Perry Mason mystery in which the heroic lawyer was not only wrong but the murder was also never solved.

Therefore, science really is just *organized common sense*. That is what a prominent scientist and friend of Charles Darwin, Thomas Henry Huxley, called it over a century and a half ago. Scientific common sense is different from ordinary common sense, however, because scientists use the hypothesis-testing approach in a disciplined fashion. For many people, just thinking of a hypothesis is good enough. But to a scientist, a hypothesis is not the end but the beginning. A scientist must test the hypothesis before accepting it.

If science is just organized common sense, one might think that science must be a very simple process. Just test creative hypotheses and draw conclusions from them. But the human mind is incapable of seeing the world objectively. The human mind is the plaything of illusions, sometimes of delusions. In this way, the mind of a scientist is no different from the mind of anyone else. A large part of the work of scientists is the attempt to recognize biases and illusions and to make corrections for them. If anything, this takes more time, effort, and resources than any other part of the scientific endeavor. That's what the next section of the book is about.

DO IT YOURSELF!

Think of something that you believe and is based at least partly on scientific information. Turn it into a hypothesis. Also specify a null hypothesis. Try to make your hypothesis falsifiable—that is, what observations could you make that might possibly show your belief to be wrong?

CHAPTER 2

SCIENCE AND FICTION: ORGANIZED COMMON SENSE AND ORGANIZED CREATIVITY

T he scientific way of thinking is not just something that scientists do. As I explained in the previous chapter, auto mechanics use the scientific method. But let me give you another example: the process of writing fiction resembles the process of science. I am a scientist, a science writer, and a fiction writer. I'm no expert on fiction, but I at least know the difference between F. Scott Momaday and N. Scott Fitzgerald. As scientist and novelist C. P. Snow asked, what could be more different than science and the arts?[1] But they aren't as different as you might think.

A COUPLE OF CLASSICAL EXAMPLES

Scientific thinking is a specialized and purified form of one of the deepest human urges: the desire to solve a mystery. The whole detective fiction genre exists because of this urge. Sherlock Holmes was famous for employing the scientific method to solve crimes. But it didn't start with him.

According to the apocryphal biblical book Bel and the Dragon (in the Catholic, but not the Protestant, Bible), a Jewish hero named Daniel defied the pagan priests in the court of the Persian king. Daniel refused to worship the pagan god Bel or offer sacrifices to him, to which the conquered Israelites were required to render worship. The priests got people to believe them on the basis of authority—they had the authority of the king behind them—and emotion, expressed through a gleaming golden altar that struck awe into the hearts of observers. But they also claimed to have proof that Bel was the great god. Their proof was that, every night, they would leave bread and meat and wine at the altar, and the next morning, the food would be gone—thus proving that Bel had eaten it. What does it matter, they said, if nobody ever *sees* Bel or hears him? Bel eats—isn't that enough to prove that he is real?

Well, Daniel wasn't buying it. He believed not in local gods who could be in a good mood or a bad mood depending on what the god had just eaten, but he believed in one great God who ruled the heavens and the earth. So, he challenged the priests to a scientific test. He wanted them to prove to him that Bel really did eat the food that was left at the altar. The king was interested in this also. The stakes were high. Daniel said that if Bel ate the food, then he should be executed, but if Bel did not, then the priests should be executed. The king and priests agreed to the challenge.

The priests prepared the food and left it at the altar. Then they were ready to close the door to the sanctuary, lock it, and seal it with the king's seal. But just before they all left the sanctuary, Daniel scattered some flour around on the floor.

The next morning, the king found his seal unbroken, and he, Daniel, and the priests entered the sanctuary. Lo, the food was gone! The priests were ready to celebrate their victory and Daniel's death. But Daniel told them to hang on a moment. He asked if they saw what he'd seen on the floor. He pointed to footprints that had been left in the flour. The footprints led to a trapdoor through which the priests had entered to take the food and eat it themselves, not seeing the flour in the darkness. The king ordered the immediate execution of the priests, and Daniel became an important official in the government.

This ancient story depicts a primitive battle of scientific hypotheses. The priests tested the hypothesis that "Bel ate the food" by having the front door locked and sealed. But Daniel tested the hypothesis that "the priests ate the food" by putting flour on the floor.

Another example of science in literature is found in Shakespeare's *Hamlet*. The ghost of Hamlet's father, the erstwhile king, had just appeared to the young prince on a dark castle turret at night. The ghost informed him that he had been murdered by the current king by having poison poured into his ears, very much like the plot of a play called *The*

Murder of Gonzago. The ghost asked Hamlet to take revenge. Hamlet found himself in an epistemological dilemma. How would he be able to convince anyone—or even himself—that he had really seen his father's ghost and not merely had a hallucination? He needed external evidence. But how do you get evidence of poisoning (in the days before chemistry) on a decomposed corpse?

Then one day a troupe of traveling actors stopped by the castle. A brilliant thought occurred to the young scientist-prince. Have the actors perform *The Murder of Gonzago*. Hamlet would watch the king to see if the murder scene provoked any sort of response from him. If the king showed no response, then he was probably innocent and the ghost was just a delusion. But if the king had a guilty conscience, he would display some kind of response. Hamlet would watch the king just as a police investigator watches a suspect during questioning, looking for evidence of guilty reactions. As Hamlet said, "The play's the thing wherein I'll catch the conscience of the king." In fact, the king did react, and Hamlet concluded that he had really seen the ghost of his father. Complications ensue.

SIMPLIFIED MODELS OF REALITY

The similarity between science and fiction might seem patently ridiculous, both to PhDs and to MFAs. Some scientists dismiss the speculations of their peers as being "okay, if you like fiction." They may smugly consider this to be the ultimate put-down. They may think of creative fiction writers as children playing around, unconstrained by the adult world of facts. In their turn, many creative writers practically ignore science, considering it to be a dead and soul-less model of human experience. I've lost count of the number of stories I have read with fundamental scientific errors. I have read the works of two fiction writers who assumed that sound travels faster than light. But some people straddle the two worlds. Examples include George Gaylord Simpson and Edward O. Wilson, scientists who wrote novels,[2] and John Updike, a novelist who was enthusiastic about science.[3] One of the major examples of a mind that explores both scientific and literary worlds is Alan Lightman, who has made major contributions to physics and has written, in addition to many delightful short stories, important novels such as *Ghost* and *Mr g.*[4] Though his novels include a lot of scientific material, Lightman's short stories often do not. Two of the most famous science writers, Carl Zimmer and David Quammen, began their writing careers in the world of fiction, not science. Maybe that is why they are so good at detecting the narrative structure within science.

Consider what both scientists and writers must do. They both have to explain something. And to do so, they need to have hypotheses (usually unspoken ones in fiction) to create simplified models of reality, then test them.

Where do hypotheses come from in fiction? Just as in science, fictional hypotheses come from things you have experienced, places you have been to, and things you know about. I, as a native, can easily recognize a story set in Oklahoma and written by someone who has not lived there. And that is why anyone who generates hypotheses—whether a scientist or a novelist—needs to get up out of the armchair and experience the world.

Fiction also uses simplified models of reality. Lots and lots of stuff happen in real life, and lots and lots of stuff happen without warning. But in a novel, everything that happens must advance the plot and must be foreshadowed. In the real world, things sometimes happen for no apparent reason. But in fiction, this is a sign of failure. Science is the same way. The universe, with all its gritty and squishy components, is very complicated. Scientific research does not simply record what happens, and say, "Look what happened." Science creates models of reality whereby the processes that operate within nature can be understood.

We carry out and read about scientific research, and write and read novels, to understand the world better. This can only be done with models of reality that are trimmed down so that their workings can be understood. A story has to be a comprehensible model of reality, rather than just a chunk of reality. It has to be a focused *refraction* of reality, not a *reflection*. Holding the mirror up to nature is not what scientists or fiction writers do. They hold the lens up to the light of reality, splitting its beams into colorful components for analysis, and focusing it into understanding.

The writer of fiction also tests hypotheses. He may create a character and then run a simulation to see how that character will react to circumstances in his or her life. The writer is *running an experiment with the character*, only the experiment is inside the writer's head. Experiments, both the scientist's and the writer's, have to pass the reality test: my hypothesis about roots has to correctly predict how actual roots will grow, and a writer's characters must behave in a way that they would in *a* real world, whether the one we live in or a different one like Middle Earth. Not just anything goes in fiction, any more than it does in science. I have said concerning some of my failed hypotheses, as I have said about some of my failed characters and plots, "This isn't working." Even if it is on a planet in a galaxy far, far away, it still has to "work."

Stephen King's *Misery* is a novel in which a famous author was held captive out in the country by an unhinged nurse.[5] The nurse did not like the way the author had allowed his main character to die at the end of his previous novel. And the nurse had her ways to get the author to write a new book to fix this problem. So the first thing the author did was to start a new book that pretended that the main character had simply not died. The nurse flew into a frenzy. She said it wasn't fair. He had to fit in the new plot with the previous plot! The author was not free to just make up whatever he pleased; he had to make it *work*. So the author came up with a plot in which the main character had gone into a coma and not actually died. The author's life depended on the success of this new plot. You can't just have a completely unanticipated force or character come in at the end to rescue a tangled plot. In ancient plays, gods would rescue people, a process that classical scholars called *deus ex machina*, God from outside of the machine. Euripides used it a lot, for example when he had Hercules show up and rescue a character just before she was going to be killed. Apparently, machines were literally involved, since the unexpected rescuer was lowered onto the stage with a crane. Scientists deplore deus ex machina even more than the nurse in King's novel did. We have to fit our explanations into a coherent network of cause and effect.

Let us consider a hypothesis that can probably never be tested scientifically, that humans living on Mars would eventually evolve larger brains. In order to test this hypothesis, you would need to have several generations of humans living on Mars. This will not happen soon, if ever, if only because it is too expensive to establish such colonies with a large enough sample size of people. The people would need earthlike atmospheric and temperature conditions, but a Mars colony would have much less gravity. One of the reasons that humans do not have larger heads is that our necks could not hold up such large heads. On Earth, that is. But maybe on Mars a thin neck could hold up a big head. Big heads would also cause a problem with the birth process; a fetus with a big head would not be able to fit through the birth canal. But humans capable of colonizing Mars would also be capable of making all deliveries caesarean. This idea can only be explored by speculation,[6] and fiction is perfectly suited to this kind of inquiry.

Fiction may also provide us with brilliant examples of bias, invalidity, and how to avoid them. If you want practice in logical thinking and in recognizing invalid assumptions and biases, you can hardly do better than to watch reruns of *Perry Mason*, a television show based on Erle Stanley Gardner novels, which ran from 1957 to 1966. The character of Perry Mason, a heroic defense attorney played by Raymond Burr, encapsulated much of what people find most attractive about the scientific method. He had a superlative ability to draw logical conclusions from evidence. But at the same time, a passion for justice bubbled underneath his logical surface. The other characters were passionate also, especially the Los Angeles County district attorney Hamilton Burger, played by William Talman. But while Burger's passion emerged very early in each episode, causing him to jump to conclusions and accuse an innocent person of murder, Mason reserved his passion for later, *after* he had carefully examined all of the alternative hypotheses. He was extremely creative, sometimes unbelievably so, in generating these alternative hypotheses. Finally, Mason frequently set up experiments, or at least simulations, within the courtroom—which Burger invariably denounced as "courtroom shenanigans." But these experiments demonstrated before the amazed eyes of all the spectators that the assumptions under which they had been working were impossible. He usually cornered the guilty party right about at the forty-five-minute mark, and only then did he pour out his passion.

In each *Perry Mason* episode, everyone was biased against the person who was initially charged with committing the murder, often because the defendant was an outsider, or less glamorous, or less rich, or had a past criminal record. Even Mason would often begin with such a bias but then question it. Like a scientist, Mason forced himself to examine alternative explanations, which allowed him to notice crucial details that others had overlooked. In one episode, a woman called Mason to report that she had been poisoned, *then hung up.*[7] The people's bias against her less glamorous cousin, the defendant, led them to overlook the fact that, at the scene of the (nonlethal) poisoning, the telephone receiver was still on the floor—everyone except Mason, of course. The woman had faked her own poisoning. Another example of Perry Mason avoiding bias is that he looked no less favorably upon poor clients than on rich ones, as when in another episode he took on as a client a woman who could pay him only thirty-eight cents over a senator who offered him ten grand.[8]

But we cannot let hypotheses limit our vision, either. Scientists, like fiction writers, have to be open to surprises. We cannot act as if we know everything. My plants and my fiction plots have surprising twists of events. We have to be willing to revise our hypotheses as new information becomes available. While working on a novel, the writer may realize that the plot demands that one of the main characters has to die. But the writer has to let this character die. That's what happens in science, when we sometimes have to give up a cherished hypothesis.

The biggest story of all, of course, is the history of the universe. Some scientists, such as Ursula Goodenough and myself, have written books about this story.[9] Neither book became famous, partly because we humans are much more interested in ourselves than in the universe.

It is true that scientists tend to be very intelligent people. We scientists have to keep a lot of facts straight in our minds as we propose our hypotheses and design our experiments and interpret the results. But the same is true of many other people. A mechanic has to keep a lot of information about cars in her brain. It is not the level of intelligence or detail that makes science different from casual thought; it is the discipline of hypothesis testing—a discipline that can be found, sometimes more and sometimes less, in other modes of thinking as well.

DO IT YOURSELF!

Think of something that you would like to know, and for which a scientific hypothesis is possible, but which could only be (realistically) tested by means of fiction.

CHAPTER 3

EXPERIMENTING WITH A MOUNTAIN

Scientists creatively test hypotheses. But in the natural world, many things are happening all at once. This can create confusion during the test of the hypothesis. Therefore, scientists prefer, when possible, to perform experiments in which they control all of the factors except the one that they are investigating. That is, scientists conduct experiments for the same reason that Alfred Hitchcock used indoor sets. He could not control the wind outside, but on the indoor set he could generate just the right amount of wind with a fan.

It is not, of course, always possible to perform an experiment. But scientists have poured a great deal of creativity into designing experiments that test hypotheses that previously had been investigated only by taking measurements from the uncontrolled natural world. Scientists do not start out by assuming that an experiment would be too big, expensive, and complex. They might have to conclude this, but they should not assume it at the outset. As this chapter will illustrate, scientists sometimes show breathtaking or even absurd creativity to make uncontrolled hypothesis testing into a controlled experiment. Scientists often do experiments in laboratories, since it is easier to regulate conditions there, but many experiments are performed outdoors also—so long as they have controls.

THE CONTROL

Setting up the appropriate control is one of the most important aspects of a scientific experiment. And it was for doing so that Francesco Redi has earned the credit of carrying out one of the first scientific experiments, the results of which he published in 1668. As strange as it may seem to us today, back then people thought that meat spontaneously generated maggots as it rotted. Moreover, nobody seemed to know that maggots turned into flies. Redi hypothesized that maggots were baby flies and that they grew from eggs that flies laid upon rotting meat. Redi placed meat in glass flasks and let it rot. He covered some of the flasks with gauze, which kept out the flies but otherwise did little to alter the conditions inside the flask. These were the controls (without flies). The open flasks, which flies visited, were the treatment. The meat in the control flasks rotted but produced no maggots. Redi also fished out some maggots and let them develop into flies. By means of this experiment that would be considered too simplistic (and disgusting) for a science fair project today, Francesco Redi invented the art of the experiment and demolished millennia of theory (spontaneous generation). Centuries later, Louis Pasteur came up with an even more ingenious control for his experiment that demonstrated that microbes cause decomposition. One of the glass flasks he used for his controls, a flask with broth but with no bacteria, is still on display in a museum—and the broth has still not rotted.

Redi's control—the flasks with gauze—had to be as similar as possible to his experimental flasks. If he had, for example, put the control meat in a flask covered with gauze, but left the experimental meat out on a plate, the conditions might not have been comparable; the meat out on the plate would have, for example, dried out at least partway. But in Redi's experiment the treatment and the control were as identical as possible, except for the presence of the gauze and therefore of the flies.

In designing an experiment, it is essential to get the right control, if there is one. Suppose you want to determine whether acupuncture is an effective kind of therapy. Your control would not simply be someone who did not receive acupuncture to compare with someone who did.

The appropriate control might be for a person to receive acupuncture at random locations on the body (avoiding, of course, certain extremely sensitive spots) rather than at the locations that acupuncturists claim to be significant (the so-called acupoints). A completely non-punctured control would not include the effects of the punctures themselves on the skin, the immune system, etc.

In some cases, it is not possible to set up a control. Some of my research involved plant growth patterns in response to light intensity. My hypothesis was that plants that grow in dim light produce relatively bigger leaves and relatively smaller roots than plants that grow in bright light. Sounds simple enough. But what would the control be? You might guess that the control would be darkness. But using darkness as a control would be misleading. In darkness, many plants grow in an abnormal way: they are white, leafless, and stringy. (Actually, these characteristics are abnormal only for aboveground growth; they are normal for a plant that is growing underground, which is the only place in nature that a seedling is likely to encounter prolonged darkness.) Therefore, instead of an actual control, I just had two treatments: bright light and dim light.[1]

But even this was a little complicated. It is impossible to increase the intensity of the light without also altering other environmental variables. In bright light, leaves become warmer because they absorb more light, and some of this heat diffuses into the air, making the air warmer as well. Warmer air, in turn, has a lower relative humidity—that is, it is drier. So just by increasing the light intensity, I have altered three variables, not just one: the "high light" plants also experience higher temperature and drier air. And there is no way to get around this problem. You could pump cool air through the high light chamber, thus cooling the leaves—but if you do so, then the air temperature is different in the two treatments. The way I dealt with this problem was to simply state that, under natural conditions, sunny conditions were warmer, just as in the experiment, so don't worry about it.

ABSURDLY CREATIVE EXPERIMENTS, PART ONE: SPIDERS

Sometimes you need some real creativity to come up with an appropriate control. In one study, researchers found that when grasshoppers encountered spiders, their metabolism changed.[2] When frightened by spiders, the grasshoppers craved (if we can use this word for grasshoppers) carbs, and their stress hormones caused their cells to break down proteins to make sugar. They ate less protein (which has nitrogen atoms) and more carbs (which do not). The reduction in nitrogen from the bodies of the grasshoppers had such a pronounced effect on the whole food chain that it even reduced the rate at which bacteria decomposed leaf litter in a prairie. This occurred because the grasshopper droppings contained less nitrogen. Nitrogen promotes bacterial decomposition. But was the grasshopper stress due to the spiders eating grasshoppers, or to grasshoppers being scared of spiders? The appropriate control would be the presence of spiders that cannot actually eat the grasshoppers. So, the researchers exposed the grasshoppers to spiders *whose mouthparts had been glued shut.* Instead of testing hypotheses on grasshoppers, spiders, and wild prairies, these scientists used an ingenious experimental setup in which spiders could scare the living daylights out of the grasshoppers without harming them. Hold onto your hats: this is just the first of several astonishingly creative experiments I will tell you about.

ABSURDLY CREATIVE EXPERIMENTS, PART TWO: FLOWERS AND HUMMINGBIRDS

Another good example of turning observations into experiments comes from the world of pollination.

First, a little background information about pollination. When a pollinator visits a flower, such as a hummingbird visiting a wild tobacco flower, there is a complex interplay of self-interest. The flowers are not just being nice to the pollinators by offering them nectar, nor are the pollinators just being nice to the flowers by carrying their pollen around.

If the flower could get the pollinator to carry its pollen around without offering it any reward (such as nectar) and without having to advertise (for example with bright petals), that flower could in theory produce more seeds by lowering its operating costs. Actually, the nectar itself is an advertisement, not just a reward. The sugar in the nectar is a caloric reward, but the nectar also has a volatile scent chemical, benzaldehyde, which the hummingbirds can smell from some distance away. But producing nectar and petals is part of the cost of doing business if you are a wild tobacco flower. Similarly, if the hummingbird could just sit and drink nectar instead of beating its wings dozens of times per second, its operating costs would be reduced and it would have more resources left over with which to produce offspring. But, for a hummingbird, flying around is part of the cost of doing business.

But how do the tobacco flowers keep the hummingbirds from simply perching on and satiating their hunger from a single flower—that is, without cross-pollinating the flowers? The tobacco plants have two main ways of doing this. First, each flower only produces a little bit of nectar. In order to get enough food, the bird has to fly from one flower to another, probably on a different plant. Second—and this is where it gets surprising—the flowers produce nectar with bitter chemicals in it. In wild tobacco flowers, the bitter chemical is nicotine. It is possible that they produce nicotine in order to irritate the birds, which keep flying around looking for some better-tasting nectar. This leads to the hypothesis that the flowers that get pollinated the most are those that produce *both* benzaldehyde (which attracts hummingbirds) and nicotine (which irritates them).

The "natural" way to test the hypothesis would be to let the flowers "decide" how much benzaldehyde and nicotine to produce. The researcher could use a little syringe to sample the nectar of each flower (out of a sample size of, say, a hundred) and measure the benzaldehyde and nicotine concentration in the nectar of each of those flowers. And then the researcher could measure floral success in two ways. First, she could keep track of how many hummingbirds visited each of the flowers and how long they stayed, thus measuring pollination success, and second, she could come back later and see how many seeds each flower produced, thus measuring reproductive success. Researchers might test the hypothesis that flowers that produced nectar with the most benzaldehyde *and* the most nicotine would be more successful than flowers that produced less benzaldehyde *or* less nicotine. This is not an experiment, since the researchers would have to just accept whatever amounts of benzaldehyde and nicotine each flower happened to produce.

But one group of researchers *used genetic engineering* to produce four groups of plants:[3]

- Plants whose nectar contained both benzaldehyde and nicotine
- Plants whose nectar contained benzaldehyde but no nicotine
- Plants whose nectar contained nicotine but no benzaldehyde
- Plants whose nectar contained neither

The plants were exactly alike except for their benzaldehyde and nicotine levels. By doing this, the researchers turned their study into an experiment, in which the fourth plant group is the control. The researchers were in charge of which flowers had which chemicals in their nectar, rather than being at the mercy of whatever the nectar of the natural flowers might have had in it.

But how could the researchers know that their results were not the by-product of the process of genetic engineering itself? Maybe the genetic engineering made some kind of change other than benzaldehyde and nicotine in the flowers to which the hummingbirds were responding. In the above four conditions, only the first represents the natural plants; the other three are genetically modified plants. But this is not what the researchers did. They genetically modified the first set of plants also: they used biotechnology to remove both the benzaldehyde and nicotine genes *and then put them back*. Had they not done so, their experiment would have been flawed, just as if Francesco Redi had left his experimental meat out on a plate while keeping his control meat inside a jar. All four treatments have to be *as alike as possible* except for the benzaldehyde and the nicotine. The experimenters found that the process of genetic engineering itself did not alter the experimental results.

ABSURDLY CREATIVE EXPERIMENTS, PART THREE: ANTS

Here is one of my favorite examples of turning an observational study into an experimental one. Three European scientists published a paper in 2006.[4] They used the typical dispassionate and sterile form of writing that prevails in the scientific literature, except in their title: "The Ant Odometer: Stepping on Stilts and Stumps." With a title like that, how can you resist reading the paper?

Ants, as I will discuss in a later chapter, appear to have a sinister collective intelligence, but actually they do not. Each ant is just following simple rules of behavior. How can we humans, biased as we are to impute intelligence to other organisms, figure out what these rules are? Therein lies an interesting tale, for later.

Now, the three scientists asked a very simple question. When an ant goes home, how does it know how far to go? Remember, it has to be something really simple for these small-brained creatures.

The scientists considered two possibilities. One was the "energy hypothesis," which says that the ants would walk homeward for as long as it takes for them to use up a certain amount of their energy reserves. The other was the "odometer hypothesis," which says that the ants were integrating the number of steps. The authors quickly dismissed the energy hypothesis because they found that the ants walked home the same distance regardless of whether they were carrying a burden (therefore using more metabolic energy) or not. That left the odometer hypothesis of step integration. "Step integration" does not mean that the ants were actually counting the number of steps, since only humans can count higher than six or seven, and most animals cannot count at all. Instead what probably happens is that for each step the ants take, a certain amount of a kind of molecule builds up or is depleted in their bodies, and when just the right amount of that chemical has built up or has been depleted, the ants are home, or should be.

But eliminating one of two hypotheses does not prove that the surviving hypothesis is correct. Suppose that some other hypothesis, not thought of by the scientists, is correct? Accordingly, the scientists sought experimental confirmation of the odometer hypothesis. Their experimental test was simple and clever. They hypothesized that ants with short legs would not walk far enough to get back home (they would "undershoot" their goal) and that ants with long legs would walk too far (they would "overshoot" their goal). But naturally occurring ants with long legs and ants with short legs may be different from one another in many ways, not just in the lengths of their legs. They may be different species. They may be different ages. They may even have different roles within the colony. Instead, it was necessary for the scientists to take ants all from the same colony, which were all more or less the same size, with the same leg length, and then shorten the legs of some of them, and lengthen the legs of others.

How do you do this? First of all, you work with large ants. *Cataglyphis fortis* ants live in the Sahara and are over a centimeter long. Second, you can shorten ant legs by breaking them. Ants, like all arthropods, have segmented legs. You can break off the outermost segment and leave the others in place. The authors referred to these shortened legs as "stumps." Third, you can make the legs longer by supergluing little bristles onto the legs. The scientists used pig bristles for this purpose. The authors referred to these lengthened legs as "stilts." As you can guess, it was this clever experimental design that made the paper famous.

At first this sounds crazy. Wouldn't stilting and stumping injure the ants? If so, how far they walked would be a measure of their injury, not of their normal walking. But the scientists found that the stilted and stumped ants could walk just as well as the uninjured ants. The only way to be sure that the manipulations had no effect on the ants would have been to produce "control" ants, with normal leg length, by breaking off their legs and then gluing on bristles that would have restored the *original* leg length. The scientists did not do this. What they did do was to allow the stumped, stilted, and unmanipulated ants to find a new food source and then go home—thus learning a new route. This time, since all the ants had the opportunity to reset their step integrator to the new route, they all went straight home. They could not have done this if they were suffering from their injuries.

The results were as the scientists expected. The stump ants did not walk far enough. They stopped and walked around in confusion, short of the nest. The stilt ants walked too far, then stopped and walked around in confusion.

You are probably thinking, *Of what possible use is this kind of research?* I would like to suggest that this study helps us to understand what goes on in the mind of an individual ant, and in the collective intelligence of the colony. Notice that the stilted ants walked right past their own nest in order to overshoot it. They obviously were not thinking, watching their surroundings, and finding their way home intelligently. They were following a very simple rule: walk toward home until the step integrator chemical tells you to stop. This study, like others, has helped us to understand how ants, following a few simple rules, can do things that look creepily intelligent.

ABSURDLY CREATIVE EXPERIMENTS, PART FOUR: FRUIT FLY HANGOVERS

One of the great insights of modern science is that all organisms have evolved from one common ancestral species. Also, all animals have evolved from one ancient animal species. This is why all animals, on the genetic and cellular level at least, have a lot of similarities. This is why we can learn a lot about human physiology by studying animals—not just those, such as chimps, that are closely related to us but even fruit flies. Many of the principles of modern genetics were first discovered by scientists who studied fruit flies, which have a new generation every two weeks. Also, there are few if any ethical objections to conducting research on fruit flies.

Scientists like to work with fruit flies. This is because they are easy to raise in little plexiglass vials. The maggots eat a special chow made from dried fruit (it smells like bananas). When the pupae hatch, they can mate and lay eggs in the chow. When the flies hatch, the scientists can take them out of the vial and put them in another. In this way, scientists can control which flies mate with which other flies. Flies are apparently not choosy and will mate with whichever other flies the scientists put them with. I might mate with strange females if I knew that I was only going to live for two weeks and that I would never get out of that damned vial. Flies do not actually *know* they are only going to live two weeks, but evolution has selected them to act as if they do. In addition, the genetics of fruit flies is very well understood. But here is the most important reason scientists work with fruit flies: they in fact share with humans many genes of great interest, including some implicated in neurological disorders such as autism.[5] Of course, flies cannot be autistic, but they can have the genes that make humans autistic.

It seems to be common knowledge that, among humans, sexually frustrated males seek alcohol. But it turns out that this desire and its resulting behavior pattern are hundreds of millions of years old. Male fruit flies display this same behavior.

This is how a group of scientists found out that male flies crave alcohol when they experience sexual deprivation.[6] It turns out that female fruit flies that have just mated will forcefully reject the advances of new males. The researchers produced two groups of male flies. The males in the first group—let's call them the frustrated group, although scientists will avoid such terms for animals that may not actually have those feelings—were placed in vials with recently mated females, who rejected them. The males in the second group—the happy group—were placed in vials with not just one but also lots of receptive virgin females. More polar opposites could not be imagined, even for fruit flies: the males forced into abstinence, and the males partying in a harem.

The researchers then offered a choice to the males from the two groups: they could either eat food with, or without, added ethanol. The flies could eat the food from little tubes from which the quantity of food consumed could be easily measured. When the researchers ran the experiment, they found that the frustrated flies chose the ethanol-enhanced food more often than did the happy flies.

Now the question arises: are the flies frustrated due to sexual depri-vation or to active rejection? To answer this question, the researchers had to use female flies that neither mated with the males *nor* actively rejected them. That is, a female fly that would just sit there unrespon-sively. They did this by letting male flies encounter unmated *decapitated* female flies. (Did I mention that scientists can be absurdly creative?) It didn't seem to matter whether the males had been rejected or merely deprived; they behaved as if they were equally frustrated.

What is going on here? Apparently, there is a brain chemical called neuropeptide F (NPF), which sexual deprivation reduces but which ethanol enhances. The researchers even measured NPF levels in fly brains. The researchers published color images of the fly brains lighting up, or not, with this chemical. Humans have a similar brain chemical called neuropeptide Y. In flies, as in guys, ethanol compensates for sex, at least on the neuropeptide level.

Based on the examples I have given you here, you can be forgiven for thinking that scientists love to torture animals—breaking ant legs, decapitating flies, gluing spider mouthparts shut—but we are not all like that. I only torture plants. I pull off their leaves, rip them to pieces, grind them up, soak their little twigs in alcohol . . . oops, I'm getting carried away. It must be that apricot beer I'm drinking. Frustrated male fruit flies are buzzing around it. I thought it was the apricot essence, but it must actually be the alcohol that attracts them.

ABSURDLY CREATIVE EXPERIMENTS, PART FIVE. PLASTIC CATERPILLARS

Some ecologists wanted to know whether predators eat caterpillars in some habitats, such as tropical rain forests, more than in other habitats. The experimental design would seem to be simple: just observe how many caterpillars get eaten in each habitat, right? But it's not quite so simple. They would have to watch all day and all night to observe predator activity. Just because a caterpillar vanishes from a branch does not mean that a predator has eaten it. Moreover, there are no species of caterpillars that live in all different habitats. How could the researchers distinguish between the predation risk of caterpillars-in-general vs. any given species of caterpillar? They found a solution to all of these problems. They used fake plastic caterpillars. Of course, predators cannot eat plastic caterpillars, but they can attack them. And when they do, they leave bite marks on the plastic and the plastic caterpillar falls to the ground. The scientists then had only to go pick up the plastic caterpillars and count the number of bite marks.[7]

WHAT DO YOU DO WHEN NO EXPERIMENTS ARE POSSIBLE?

Scientists expend a great deal of effort and creativity to design experiments whenever possible. But there are of course whole fields of science in which experiments are impossible. Astronomy, for example. You pretty much just have to look at the stars, or measure them in some other way, and let it go at that. Astronomers create computer models and compare them with their observations and measurements, but astronomers cannot perform experiments with stars and planets. Astrophysicists can perform some experiments, for example with particle accelerators. But this is a long way from working experimentally with stars or black holes.

Geology is also largely nonexperimental. For example, I mentioned previously that the dinosaurs became extinct sixty-five million years ago because a giant asteroid fell out of the sky. This has clearly got to be one of the best science stories ever. And it happens to be true, too. When Luis W. Alvarez (a physicist) and his son Walter Alvarez (a geologist) first announced it, most scientists resisted it. But the Alvarez team found evidence that eventually proved convincing.[8]

To test the asteroid hypothesis, the scientists looked for some chemical element, deposited at the time of the putative asteroid impact, which is common in asteroids but rare on Earth. There in fact is such an element: iridium. There is a thin layer of clay between the dinosaur-era rocks and the post-extinction rocks that is rich in iridium; in the other rocks, iridium is rare. No experiments were possible, but the data were convincing nonetheless.

ABSURDLY LARGE EXPERIMENTS

Some hypotheses, for example the ones related to asteroids and black holes, are simply too big to test experimentally. But don't just assume that if an experiment is not easy it cannot be done. Back in the 1960s some ecologists wanted to test the hypothesis that a forest on a mountainside will slow down the flow of rainwater, and thus prevent soil erosion and flooding at the base of the slope. Makes sense, but proving it is another matter. What they did, with appropriate permission and help, was to cut down the trees on part of a mountain slope and leave the forest intact on other parts of the slope. At the base of the slope, they could measure the amount of mud, minerals, and water that ran off that particular part of the mountain whenever it rained.[9] They had to make sure that the clear-cut and the intact slopes were otherwise as similar to one another as possible. They also had to make sure there was a good layer of rock under the slope so that the water would flow downhill rather than down into the earth. These scientists, therefore, experimented on a mountain.

Other scientists experiment with whole forests. As mentioned in other chapters, humankind is pouring carbon dioxide into the air with our industry, transportation, and energy production, causing global warming. The scientists want to know whether this extra carbon dioxide will cause the trees of the forest to grow more. If it does, then maybe the trees can absorb the carbon dioxide and help stabilize atmospheric carbon dioxide and prevent global warming. And if they do so, by how much? These scientists have set up Stonehenge-like rings of towers that release carbon dioxide inside the ring, making the air inside the ring more carbon-rich than the air outside. The inside (the experiment) and the outside (the control) are otherwise nearly identical. The towers measure wind speed and carbon dioxide levels, and release carbon dioxide at just the right times and places. Back in the old days, scientists ran these experiments inside of greenhouses. I managed some of these early experiments. But greenhouses tend to be hotter and more humid than the real world. If you really want to know how a forest responds to elevated atmospheric carbon, you have to experiment on the forest, not on a few trees in a greenhouse.[10] Similar experiments have been done on farms.[11] The news is not good: neither forests nor farms can remove enough carbon from the air to prevent catastrophic global warming.

It is more than just unfortunate that we humans are "experimenting on" Earth today by pouring carbon dioxide into the air. This is not really an experiment. First, there is no control—there is no backup Earth on which the experiment is not being done. There is one globally mixed atmosphere. They are just playing around with Earth the way little kids "experiment" on flies by pulling their legs and wings off. Things therefore do not look good for what the late Stephen Schneider, an atmospheric scientist and one who believed the public needed to hear about global warming, called "Laboratory Earth."[12]

CHAPTER 4

WRIGHT AND RONG

In life, there are lots of ways of being wrong. But in science, there are just two ways of being wrong.

The first way of being wrong is to accept a *false positive*—to conclude that the hypothesis is correct when, in fact, it is not. The second way of being wrong is to accept a *false negative*—to conclude that the hypothesis is wrong when, in fact, it is correct.

Both kinds of errors—false positives and false negatives—are wrong, but in general a false positive is worse than a false negative. Here's why. Consider an experiment that tests the effectiveness of a drug. A false negative would conclude that there is no evidence that the drug works when, in fact, it does. The only problem this causes is that the researchers may redesign the experiment, or reformulate the drug, and try again. It results in a delay in getting the drug to market. This can, of course, cause scientists to waste time and money, which is why Stuart Firestein has called for the publication of "failed" experiments.[1]

But suppose that a scientist is not willing to give up, quite yet, on the hypothesis. A false negative can then inspire a redesign of the research. In the plant-root experiment that I described previously, my hypothesis was not confirmed (that is, the data agreed with the null hypothesis) the first time my students and I gave it a try. We used bean plants in this first experiment. It turns out that there were two reasons why the bean roots did not preferentially proliferate in the good soil. First, bean seeds are pretty big and have a lot of nutrients stored up in them, so perhaps it was simply not important for the roots to seek out more nutrients. Second, beans, like many other leguminous plants, form mutualistic partnerships with bacteria. The bacteria act as little nitrogen fertilizer factories. Therefore, bean seedlings do not need to get nitrogen fertilizer from the soil. They still need to get phosphorus from the soil, but maybe the seed already has enough phosphorus in it. So, we redid the experiment, using sunflowers instead of beans. Sunflower seeds are smaller, and sunflowers do not have mutualistic fertilizer-creating bacteria in their roots. This second time around, we confirmed the hypothesis and dismissed the null hypothesis. The false negative we got from the first experiment merely delayed us in reaching the conclusion—and actually taught us something interesting along the way.

But a false positive—for example, believing that a drug is effective when in fact it is not—would make all the researchers jump up and say, "Eureka!" Then the company would allocate a lot of funds to the drug, only to find out, many research hours and perhaps millions of dollars later, that the answer should have been "phooey" instead of "eureka." As I mentioned previously, I have identified a plant extract with potential medical applications. I am an amateur at such research. I provided a small pharmaceutical corporation with information about my research. They were interested and began to investigate the plant extract that I sent them. The first thing the company did was to repeat all of my measurements, using a sample that I sent them and their own standardized methods rather than my amateur ones, before taking the research any further, to protect themselves against a false positive. They found out I was right. Any corporation that did not carefully guard itself against false positives would be quickly out of business. (The small corporation has since been bought out by a larger corporation, which closed down this research project.)

You can never, ever, ever be absolutely sure whether you are right or wrong in science (and, I believe, in any other aspect of life). To insist on absolute certainty would be paralyzing. So, scientists have decided to take a slight risk. Scientists have universally decided to take a *5 percent risk* of being wrong. That is, they will tolerate a probability of accepting a false positive or a false negative that is 5 percent or less. That is, they accept twenty-to-one odds of being right but not less. This is arbitrary, but you have to have some kind of standard to work from. This leads to the strange situation in which a team that achieves an error probability of 4 percent celebrates while a team that achieves a probability of 6 percent feels defeated. Anyway, just remember that scientists have agreed, as one of the rules of membership in the community of scholars, to accept an error rate of 5 percent but no more. (The 6 percent group might try again, maybe increasing their sample size a little or refining their techniques, in the hopes of reaching the 5 percent level. That is, assuming their failure was a false negative.) Scientists use statistical programs to calculate these probabilities.

In some cases, the 5 percent standard may not be good enough. An increasing number of medical researchers believe that, when patients' lives are on the line, a 5 percent risk is too high. Some of them are calling for ten times as much stringency (a half percent, rather than 5 percent, risk).[2]

Just because the results of one experiment have only a 5 percent chance of being wrong does not mean that, if you repeated the experiment all over again, you would get results that were significant at that 5 percent level. You might end up with a 6 percent chance, which is "not significant," or you could end up with a 4 percent chance. That is, your results might, by chance, be better or worse. This is another reason that many scientists prefer a more stringent threshold of significance than just 0.05.

Scientists really, really enjoy having their research results fall below the 5 percent threshold. In fact, their jobs may depend on it. If a young scientist conducts research that keeps "failing" and that produces null results, the scientist may soon find herself unemployed, whether in academia or in private research labs such as those run by pharmaceutical corporations. Null results usually also close the door to competitive research funding, especially from the federal government (the National Science Foundation and the National Institute of Health being two major examples). This has led to two unfortunate consequences (some call them crises), especially in the high-stakes world of biomedical research.

The first consequence is that fewer and fewer scientists are willing to undertake risky research—that is, research with big ideas but a high risk of failure. As a result, most scientific research consists of small alterations of existing research. This is called "safe science" because it is very similar to research that has already been proven to work. But it is the big ideas, the risky science, that is most likely to lead to major breakthroughs. Roberta Ness refers to this as the "creativity crisis."[3] Creativity is risky, and many scientists stay away from it.

The second consequence is that scientists feel a great amount of pressure to make their results "work." This leads many of them, consciously or unconsciously, to cut corners and reach "eureka!" conclusions when the data simply do not justify such a conclusion. The result, according to Richard Harris, is often faulty medical research for which billions of dollars have been spent.[4] Many of these errors occur because of bias, meaning that scientists see what they expect to see rather than what is actually there, or because of invalidity, meaning that the results cannot be applied outside of the laboratory. Avoiding bias and ensuring validity are such important parts of the scientific process—and of common sense in daily life—that I devote much of the next section to them.

It is also possible that a scientific investigation might find a *statistically* significant result—avoiding both false positives and false negatives—but the result is not *scientifically* significant. A drug might enhance health by 1 percent, and this 1 percent might be statistically significant, but it is not worth the cost of drug development.

In science, as in all of life, there are no ways of being absolutely certain. Scientists, at least, can calculate the risks they are up against.

DO IT YOURSELF!

Think of a hypothesis you would like to know about and how you might investigate it. What if you made a mistake and ended up with a false positive? Or what if you ended up with a false negative? Tell, for your hypothesis, which kind of error would be worse.

SECTION II

LEGACY OF AN APE'S BRAIN

As I noted previously, the human brain did not evolve to reason; it evolved to rationalize. We did not evolve to discover the truth unless that truth was directly beneficial to our evolutionary interests. Therefore, science does not come naturally to us. In particular, our minds are warped in many and curious ways by bias and invalidity. That is what this section of the book is about. But I will show you that the joyous discipline of science can allow us to come as close as possible to truth as our ape brains will allow us.

The human mind looks for patterns in the world. If it finds no patterns, it imagines them. This is true for all of us all the time. We therefore "see" patterns, and sometimes objects, that are not really there. And the human mind can be blind to things that it does not expect to encounter. Even when a pattern is shown to us, or an object is right in front of us, we may not be able to see it. Scientists, therefore, must take special measures to help them avoid these mental lapses.

The discrepancy between the way the world actually is and the way we perceive it is called *bias*. There are many kinds of bias. David Chavalarias and John P. A. Ioannidis claim that there are 235 kinds of bias in biomedical research.[1] In this book, we will encounter a few of them. They include the bias of trusting our senses, the linear bias, the categorical bias, the bias of assuming that correlation is causation, and the bias of agency. We often used the word "biased" as an epithet for someone who is too careless or heartless to consider viewpoints other than his or her own. But bias does not have to be deliberate or even conscious. Bias is something that we all have. It starts with the very act of measurement itself. But it certainly does not end there.

Furthermore, we can think that we have reached a conclusion based on a set of observations, only to discover that our observations were *invalid*. A set of measurements is valid if it tells us what we think it tells us.

This section describes examples of how wrong we can be, through bias and invalidity, without even knowing it, unless we embrace the discipline of science.

CHAPTER 5

A WORLD OF ILLUSION

Being a synesthete must be quite a trip. Synesthesia means that feelings (aesthetics) come together (syn-). Synesthetes experience sensory inputs in multiple ways rather than in just one. For example, they may taste musical chords and see colors in response to music. I have known three people with synesthesia, all of whom are musicians. One of them astonished his non-synesthete friends once by saying after a concert that the music "tasted so good." He was surprised that they could not taste the music; he thought it was just an ordinary way of experiencing the world. The other two synesthetes saw colors when they heard music: to one, brass instruments made the world turn red, and to the other, beautiful music made the world turn purple. I'm almost envious. Some people put themselves through considerable risks to achieve something that comes naturally to a synesthete's brain every day.

At first it might appear that synesthetes live in a world of illusion, unlike the rest of us. But, as it turns out, all humans live in a world of illusion.

The only way we know anything about the world is through our senses. Our brains are clumps of prodigiously interconnected neurons that process information that the sensory neurons deliver to them. The brain itself, floating in fluid, protected inside the skull, and fed an uninterrupted supply of food and oxygen, does not even have its own pain sensors; a patient can watch his own brain surgery reflected in a mirror. All the sensory information that the brain receives, from whatever source, is the same: nerve impulses. There is no chemical difference between a nerve impulse from the eyes or from the tongue. It is the brain that sorts the impulses out. The brain interprets the nerve impulses from the eye as vision, and the impulses from the tongue as taste. In most people, the brain keeps them distinct; in synesthetes, it does not. But in all of us, the brain *creates an illusory model of reality*. Let me give some examples.

Ready for this? You're not going to believe this. There is no such thing as color.

The photons of light have different wavelengths; some light has short and energetic wavelengths, about four thousand times the width of an atom, while other light has long and less energetic wavelengths, about seven thousand times the width of an atom. And everything in between. "Light" energy, at yet shorter wavelengths, is very powerful— such as X-rays. Long-wavelength light has less energy, for example the infrared radiation that we feel as heat. Visible light photons stimulate the three kinds of cone cells in the retina of your eye. Short-wavelength light stimulates one kind of cone cell, which then sends a message through the optic nerve to your brain. Long-wavelength light stimulates another kind. Your brain interprets the impulses from the first kind of cone cell as blue and the second as red. And everything in between. Light itself does not have color; it only has wavelengths. Color is an illusion created by your brain. That is why the *continuous* range of wavelengths found in natural light looks to us like a rainbow of *discrete* bands of color.

Illusion is not the same as *delusion*. The brain constructs for us a model of reality, using nerve impulses from sensory structures all over the body. This is an illusion. But if it were a delusion—for example, if the illusion did not include the edge of a cliff when, in reality, that edge was there— then the animal whose brain was so constructed might fall over the edge and not pass on genes to the next generation. Natural selection assures that most animals are free of delusions.

Natural selection has reinforced our emotional responses to the brain's illusory model of reality. Bright colors and sweet tastes are both associated, in our brains, with pleasure. This stimulated our primate ancestors to seek out ripe fruit, which were and are good sources of nutrition. Cats, on the other hand, cannot taste sweetness. They have no sweetness sensors. Instead, their umami sensors (the tongue sensors that respond to chemicals in meat) are more numerous than ours, and their brains associate meat with pleasure. In hummingbirds, which love nectar, an ancestral umami sensor has evolved into a sweetness sensor.[1] Natural selection has shaped the minds of each species of animal to experience the world in a way most useful to their survival and reproduction.

Do all of our minds perceive things the same way? When you see green, do you see the same thing I do? I would say that, with certain exceptions, the answer is yes because all humans are genetically very similar (we share a common African ancestor that lived around a hundred thousand years ago) and our brains are all wired similarly. One exception is that some people (mostly males) are red-green color-blind. That is, one of the classes of cone cells in their eyes is defective.

Our illusory perception of the world is illustrated not just by what we do see but also by what we don't. A lot of people do not even notice plants, even when looking directly at them. This phenomenon has been called "plant blindness."[2]

This is also why our imaginations can be almost as real as the world directly perceived. Probably no other species has imaginations quite as vivid as ours. Eagles can see better, dogs can smell more acutely, but humans have the ability to create mental models of reality—through art, religion, and science—that is totally out of the animal ballpark. Am I speculating too much to attribute the evolution of our big brains in some measure to the adaptive advantage of imagination? As Henry David Thoreau said in his 1862 essay "Autumnal Tints," "We cannot see anything until we are possessed with the idea of it and take it into our heads—and then we can hardly see anything else."

People with enhanced sensory experiences may have played an important role in human evolution. In early human societies, people who experienced the world more vividly—and who could convey that vividness to others—may often have gotten positions of social dominance. They thought, and convinced others, that they could see the spirits and talk with the gods.

Not only do different individuals see things differently from one another, but different cultures can do so also. When experimental subjects looked at a photograph, the American subjects looked at the central object in a photograph, for example a tiger, while the Asian subjects looked not just at the central object but also at the surrounding scene. This was confirmed by precise measurements of eye movements.[3]

The human brain cannot just perceive things as they are. This is not a defect but a blessing. I am not just a scientist who happens to have a rich artistic experience of the world. Science actually enriches my experience, and that is what I will write about at the end of the book. However, it does mean that human senses cannot be trusted as scientifically valid measuring devices, as I will explain further in the next chapter.

DO IT YOURSELF!

Come up with your own example of something that your senses might perceive but for which they would give you a biased representation. Use one of the senses that has not been described in this chapter.

CHAPTER 6

JUST MEASURE IT!?

The mind's inability to objectively view reality begins with even the act of measurement itself. What could be easier, we might think, than to test a hypothesis—find the things you need to measure and just measure them? But this is not at all an easy thing to do.

HUMANS AS MEASURING DEVICES

Humans have a very limited range of perception. We respond only to sensory stimuli, which provided an evolutionary advantage to our ancestors. Radio waves are passing through your body right now, but you cannot perceive them; for you, they are not stimuli. We have developed technologies that have vastly extended our range of perception. We have invented antennae that can convert radio waves into electricity, which cause a speaker to produce sound. Another example of a technological extension of perception is the microscope. Microscopes that focus light through glass lenses can allow us to see very tiny objects. However, light microscopes cannot focus an image that is smaller than the wavelengths of visible light, somewhere around a half of a micrometer. That's pretty small—a micrometer is a thousandth of a millimeter—but lots of detail in the natural world is smaller. Electron microscopes can focus beams of electrons much more accurately, so accurately, in fact, that science journals now routinely publish images of atoms! Since atoms are constantly moving, the images are of atoms in deep-frozen crystals.[1]

Within these limitations, however, we assume that our senses provide objective information about the world. We see things that are really there. We hear vibrations in the air and detect molecules in the air with our noses. Usually, this assumption is correct. But not always.

One reason is *sensory fatigue*. Our brains evolved in our animal ancestors as, among other things, a way of detecting and reacting to danger. Therefore, our brains, like those of all animals, are exquisitely sensitive not so much to environmental information as to *changes* in environmental information. We may be able to *see* an unchanging landscape, but we *notice* things that change. After a while, our minds begin to ignore some of the sensory information that comes to them. You could feel your clothes when you were putting them on, but once you have had them on for a while you forget about them. You can feel a chair when you first sit in it, but after a while you stop feeling it. When you first encounter a new scent, you notice it, but after a while you stop smelling it. The reason your brain does this is that unchanging environmental information is unlikely to be dangerous. Food that smells or tastes slightly bad might be toxic, and you need to know it right away, but after you have decided not to eat it, the information is no longer useful. The brain, in order to keep itself from sensory overload, selectively ignores old information. This can be important in measurement. Your nose is an exquisite device for detecting and distinguishing odors. But the fact that you do not smell something might mean that it is no longer there, or it might mean that your brain has started to ignore it. A mechanical odor detector can objectively indicate the presence or absence of a chemical; your nose cannot. It gets tired: sensory fatigue.

Another reason is *sensory acuity*. For any one of your senses, each part of the body has a different level of acuity. For example, all over your body, your skin has nerve endings that detect pressure. These are distinct from the nerve endings that detect pain, the ones that detect heat, and the ones that detect cold. You are wired, my friend. These nerve endings give you a sense of touch. But you have a greater density of pressure-sensitive nerve endings on your fingertips than you do, for example, on the back of your neck. You can distinguish two points of contact against your skin, and you can find and remember the location of a point of contact, much better on your fingers (even when you are not looking at them) than you can on the back of your neck. The reason should be obvious. We use our fingers to get detailed touch-related information about the world, but on the back of the neck, all we need to know is if a centipede is crawling on it. If we felt everything perfectly all the time, the result would be sensory overload. A mechanical pressure detector provides an objective and unvarying measurement of pressure; your skin does not. Just measure pressure? You *can't do it* with your skin's pressure sensors. That's why scientists use mechanical pressure sensors.

Another example involves vision. A camera can provide an objective measure of light intensity. The photographer can allow more light into a camera by opening the aperture (the adjustable hole through which light enters), and less light by restricting the aperture. In the real world, the sunlight outside can easily be a hundred times brighter than artificial light inside a room. If you keep the exposure time and the aperture the same, your photograph of the inside of a room will probably look almost black, and your photograph of the outside world will probably look almost white. You have to adjust the aperture to compensate for this. With a camera, you may have to do this intentionally. But your vision makes this adjustment without your knowledge. This occurs partly because the iris of your eye opens or closes like a camera aperture, and partly because your brain adjusts your sensitivity to light intensity. Your sense of vision may be very good, but it is not an objective way of measuring light intensity. Certainly, the inside of a room does not *look* a hundred times dimmer than the sunlit landscape outside. Modern digital cameras make these adjustments automatically. Just measure light intensity? You *can't do it* with your eyes. That is why scientists use equipment such as "quantum sensors," which convert light energy directly into electrical voltage to measure light intensity.

It gets even more complicated. If you take a photograph of an outdoor scene that is a mixture of sun and shade, either the sunny areas will be overexposed or the shady areas will be completely dark, no matter what combination of the exposure time and the aperture you use. Photographers routinely get around this problem by using an electronic flash—yes, even or especially outdoors on a sunny day—to fill in the shadows. But your sense of vision makes these adjustments automatically. Each group of retinal cells in the back of your eye connects to an optical neuron; in some very sensitive parts of the retina, each cell connects to one optical neuron. When the impulses from these neurons reach the visual processing center of your brain, the brain adjusts the intensity of each impulse. Your brain allows your sense of vision to fill in the shadows and dim down the sun. It is as if each of your visual neurons has its own aperture control. Some more advanced digital cameras can do this as well, thus making a digital camera less like an eye and more like a brain.

The evolutionary reason for this should be fairly obvious. Our animal ancestors needed to be able to see potential threats, such as predators, in the shade, and to do so quickly. Vision that adjusts quickly to changes in light intensity, and is more sensitive to light from the shade than from the sunlit landscape, spells the difference between life and death, and has done so for over six hundred million years. But this same capacity that allowed our ancestors to survive also makes the eye unable to objectively measure light intensity.

This also explains why our brains notice sudden or unusual movements more than gradual or ordinary ones. You have probably had the same experience as I: you see a movement of a leaf or bird from the corner of your eye, and you have a brief rush of preparation, as if it is a predator. In fact, your brain starts to respond to this movement even before you are consciously aware of seeing it. We are descendants of animals that detected predators on flimsy data, even when our brains responded to a false positive (assuming the predator is there when it is not). Because, you know, a false negative (assuming the predator is not there when it is) would mean the end of our genetic lineage.

WHAT DO YOU MEASURE?

All of the previous examples are of ways in which sensory bias affects our very act of perceiving the world. But suppose that you do, in fact, have a way of measuring something without bias. Maybe a ruler to measure length, for example. From that point on, you might think that all you have to do is just measure it. Suppose your hypothesis is that men are taller than women. Well, just measure the height of a man and of a woman and draw your conclusion, right? You know this is not true. The scientific method explains why.

First, consider that the hypothesis does not claim that *all* men are taller than *all* women but just implies that men, *on the average*, are taller. Second, consider that you cannot measure the heights of all men and of all women but only of a *sample* of each. A *mean* is the average of the members of a specified group. In this case, the all-inclusive group is the entire populations of men and of women, and within them there are samples. You cannot know the *population means*, which is the average height of all men or of all women, in the world and who have ever lived. You can only measure the *sample means*.

This is a problem of *validity*. A measurement is *valid* if it tells you what you really want to know. In this example, a valid sample mean must be representative of the group about which you are drawing a conclusion. A single man and a single woman are not a valid sample of the population. In order to be valid, your samples have to be larger. Your samples have to reflect the diversity within the actual population. If your population contains different races, your samples should include them, also. And if your population contains different ages, you should only make comparisons between males and females of the same age.

Once you have gotten your sample of men and of women, you get out your tape measures, or get your volunteers to stand against a wall that has measurements marked on it. (Scientists use metric measurements, such as meters and centimeters, rather than traditional measurements, such as feet and inches.) And so, you determine the height of each man and each woman in your samples. Sound simple enough?

You've probably guessed that even this is not as simple as it sounds. How precisely do you measure? To the nearest centimeter? The nearest millimeter? It depends. Measuring their heights to the nearest millimeter is not too realistic. For example, each person may have a different amount of hair. Does height include the hair, or not? How about the shoes? What happens if the person is slouching just a little bit? But so long as you measure to the nearest centimeter, rather than the nearest millimeter, it might not matter too much. If you expect the difference in height between men and women to be on the order of a few centimeters, then your precision needs to be only to the nearest centimeter. To be more accurate than that is a waste of time. Some of my students write down all the digits that show up after the decimal on their calculators and are surprised when Mr. Scientist tells them to just round it off to the nearest integer. But you don't want to use numbers that give a false sense of precision.

By the way, precision and accuracy are not the same thing. Precision refers to the fineness with which you have measured something: To the nearest meter? The nearest millimeter? The nearest nanometer? Accuracy refers to whether the digits that you have measured, within the level of precision that you have specified, are correct. We have to be careful with these concepts, whether in science or in daily life. To say that there are 7.6 billion people in the world was accurate at the time this book was written and reflects the correct level of precision: that is the number of people in the world to the nearest 0.1 billion. But if you say there are 7,600,000,000 people in the world, you have implied that there are not 7,600,000,001 people, which is not what you meant to say. Precision and accuracy are the reasons for scientific notation. If we say there are 7.6×10^9 people in the world, it is clear that only the two digits 7 and 6 are accurate.

Actually, there is yet another possible problem here. It may seem easy enough to measure the height of a person, but what about the height of a tree? You can't climb up to the tippy top of the tree and let down a tape measure. So, what do you do? You can use trigonometry, or triangulation. If you know how far you are from the tree and the angle of the top of the tree from your line of vision, you can calculate the height. (The formula is distance multiplied by the tangent of the angle.) You have to add in the height of the distance of your eyes from the ground, usually about 1.5 meters. This method is not good enough for you to worry about being more precise than just the nearest meter or so.

You might think that you would never need to use triangulation on humans. But in the nineteenth century, Charles Darwin's cousin Francis Galton was studying the body shapes of Hottentot women in Africa.[2] (These are people we today call the San.) Galton was impressed with their steatopygy—that is, the storage of fat in their buttocks. He wanted to know how wide their hips were, but he did not feel comfortable with using his tape measure on the women's bodies. So, he measured them from a safe distance and used trigonometry to calculate the width of each body!

Very well. Now, suppose you measure the heights of five men and five women and get these results:

EXAMPLE 1

	Men	Women
	171 cm	168 cm
	175 cm	166 cm
	169 cm	167 cm
	181 cm	165 cm
	177 cm	207 cm
Means	175 cm	175 cm

Figure 6-1. If your sample contains a value that is very different from the others, that "outlier" might cause the sample mean (average) to give a misleading representation of the sample. (Cartoon by Leslie Gregersen.)

The men and women in your samples have the same average height. But would you conclude that men and women, in general, have the same average height, thus rejecting your hypothesis and confirming the null hypothesis?

No. The average height of the women is inflated by the one 207 cm (six-foot-nine) woman in the sample. Aside from her, all of the women are shorter than any of the men. Such a number as 207 cm is called an *outlier* and is not considered a valid part of the sample—that is, it causes your set of measurements to not tell you what you want to know (figure 6-1). What can you do about outliers? There are statistical methods for excluding them from your samples. Or, you can just get a bigger sample so that the outlier is not such an important data point. What you *cannot* do is to just secretly throw out any number that you don't like without saying that you have done so.

Now, suppose instead that these are the numbers you got:

EXAMPLE 2

	Men	Women
	171 cm	170 cm
	175 cm	176 cm
	169 cm	168 cm
	181 cm	182 cm
	177 cm	174 cm
Means	175 cm	174 cm

In your samples, the men are on average a centimeter taller than the women. Would you then conclude that men are, in general, a centimeter taller than women, thus confirming the hypothesis?

No. You can easily see that the *variability* of the measurements *within each sample* (169 cm to 181 cm in men, and 168 cm to 182 cm in women) greatly exceeds the one-centimeter difference *between the samples*. You would strongly suspect that the difference in average height between the two samples was due to chance.

Scientists have a way of calculating just how much variability each sample has. In a sample that has no variability, all of the numbers are the same as the mean. In a sample with a lot of variability, many of the numbers are quite different from the mean. You can always calculate a mean or average for any set of numbers. But if the numbers have a lot of variability, the average or mean may be, shall we say, meaningless.

So what do you do? You sort of have a hunch that the one-centimeter difference is not meaningful. Your subconscious mind might have even done some calculations of which you are unaware and that led to this "hunch" popping into your consciousness. But to avoid bias (about which I have many more chapters of things to tell you), scientists use *statistical analysis*. Statistical analysis can tell you whether or not the two sample means differ enough to be discernible despite the inevitable variability.

When a scientist says that two sample means are *significantly* different, he or she means that there is no more than a—you guessed it—5 percent chance that they are the same. "Significantly different" does not mean "different enough to matter to me." It has a precise mathematical meaning.

This book is about how you can think like a scientist. One way you can do so, as this chapter has shown, is to base your beliefs on enough measurements from valid samples, and to be as precise as necessary without overdoing it and creating a false sense of precision. But when it comes to determining the significance of your results, I am afraid this book cannot help you. To understand even the simplest statistics probably requires a full book; I might suggest the *Cartoon Guide to Statistics*.[3] The mathematics is rather complicated. There is more than one way to measure variability in your samples—the variance, the standard deviation, the standard error, the coefficient of variation, the Gini coefficient, and others. Which one should you use, and how do you calculate it? To answer that question would require a chapter in itself. Of course, computer programs can compare sample means quickly. Feed in the numbers, and the computer may tell you "p = 0.025," which is less than 0.05, which indicates that your results are significant. But these computer programs are for sale and they are not cheap. The best you can probably do without such a program is, after getting the best samples you can get, to calculate the means and visually inspect the variation and make an intelligent hunch about whether the difference between the means is significant.

There is almost always more than one way of measuring something, or of calculating it. I learned this very early in my science career. I was a freshman taking a first-quarter calculus course from Ken Millett at the University of California, Santa Barbara. He explained that you could use integral calculus to determine the area underneath a curve. But then he said that you could also draw the curve on paper, cut it out, and weigh it. If you know the density of the paper, you can calculate the area under the curve. There's no rule that says you have to use calculus. Nor do you have to use the finest equipment. During part of my thesis work, I made some of my best measurements—not the most precise but the most useful—with a string, a nail, and nine pieces of wood.[4] (I was measuring leaf areas in wild habitats.)

It would be nice if things were so simple that a single measure would give us the answer to a question or test a hypothesis. But the world is way too complex for that.

CHAPTER 7

WE SEE LINES WHILE NATURE THROWS US CURVES

The next bias I will address is our tendency to see everything in a linear fashion when in fact most natural processes are nonlinear. Linear refers to a line. A straight line. Nonlinear usually refers to a curve. Where the distinction becomes particularly important is in understanding processes or relationships, rather than just shapes.

The linear bias made little difference to the survival and reproduction of our ancestors over evolutionary time. If it takes a day to get from one place to another, it takes two days to get twice as far. Velocity is a linear process. No problem.

But consider an object falling to the ground. As the object falls, under the influence of gravity, it starts off falling slowly, then speeds up more and more. That is, it accelerates. When an object falls, it moves slowly during the first second of its fall. But, if it falls a long way, it can be moving very, very rapidly during its last second. This has made little difference during most of our evolutionary history. If a predator, or a person running away from a predator, jumps from a tree or a cliff and subconsciously estimates his or her rate of fall in linear terms, the resulting error is very small. The only time the error would be great is if the person falls for a great distance, perhaps a hundred feet or more. The last thought to go through that person's mind is certainly not going to be *Zounds! My linear estimate of terminal velocity was wrong!* The linear error became important for understanding the trajectory of arrows or catapulted stones or planes or rockets, but those technological innovations are comparatively recent in our evolutionary history.

GROWTH CURVES

Another example of the failure of linear projections regards growth. For decades, a favorite example of a nonlinear growth curve has been the story of the frog and the lily pads. Consider a frog sitting on a lily pad in a pond. The population of lily pads doubles every day and will completely fill the pond at the end of a month (day 30). The question is, On what day is the pond *half* filled with lily pads?

The intuitive, linear answer is day 15. Half filled on day 15, filled on day 30, right? But because the doubling time of a population is a nonlinear process, the correct answer is day 29. The pond is half filled with lilies on day 29, and then the lily population doubles, completely filling the pond on day 30. Environmentalists, such as Lester Brown in his book *The Twenty-Ninth Day*,[1] love this story because it applies with chilling insight to the nonlinear growth of populations, most notably the human population.[2] Population growth can take us by surprise. A complacent (and linear-minded) frog, on day 29, could look around and croak, "Some frogs say the pond is half filled, but I say it is still half empty!" even though Frogmageddon is only a day away.

Have our brains failed us in the matter of population growth? During most of the time in which our intelligence was evolving, the answer was no. This is because most human populations had the capacity, but not the opportunity, to experience nonlinear growth. A little population explosion might start, only to be stopped in its tracks or catastrophically reversed by a plague or a conquest or a famine. Only in recent centuries have disease and famine been sufficiently suppressed to allow nonlinear population growth. As a result, the world population has more than doubled during my lifetime, even though the doubling of the world population took many hundreds of years back in the days of the Romans.

Nonlinear growth of populations helped to explain why plagues of insects occurred in ancient times (and why they occur today). But nobody in ancient times knew what caused plagues of insects—for anything they knew to the contrary, God caused them miraculously, as Moses said in the biblical book of Exodus. Ancient people seldom had any chance to estimate insect population sizes in order to project the coming of a plague because the insects frequently swarmed. A population of insects, growing nonlinearly somewhere else, could fly in over the horizon and eat up your entire crop.

Part of our problem in dealing with nonlinear growth is that it can lead us into numbers that are beyond our intuitive familiarity. We are comfortable with thinking about numbers up into the hundreds, maybe thousands. But as we get close to a million, our brains start freezing up. Imagine a thousand. Now imagine a thousand of these thousands. That's a million. And it gets worse from there. Imagine a million. Imagine a thousand millions. That's a billion. And it gets worse. How many millions in a trillion? I asked an honors biology class this question. Nobody knew. One person hazarded a guess. "Three?" he asked. The correct answer is that it is a million. A trillion is a million millions. With even honors college students not knowing how much a trillion is, how can we expect our future leaders to guide us in a world of nonlinear growth? The US national debt is $21 trillion.[3] A trillion here, a trillion there, it starts to add up to real money. People's eyes glaze over.

Population growth is nonlinear on the microscopic level as well. Bacterial populations in milk double over time. If their populations had linear growth, then milk that was just a little bit sour after one week would be twice as sour after two weeks. But because bacterial populations double, when the milk first starts to turn the least bit sour, full spoilage is only a few hours away. When bread rises, yeast populations grow nonlinearly. Ancient people did not know this, but it didn't matter much. They simply learned how to make cheese, bread, yogurt, kefir, and koumiss without analyzing the underlying growth of populations of microbes of whose existence they knew nothing about anyway.

NONLINEAR ECONOMIES OF SCALE

The linear bias might lead us to overlook the economies of scale, which are nonlinear. To run a company with two thousand employees, you do not simply do the same thing a company with twenty employees would do, but a hundred times as much. You don't just "scale up" in a linear fashion. You have to do things differently. A small company may consist of employees each of whom do a little of everything. But a large company has specialized employees each of whom can do a particular task better than a generalist could. These are examples of positive effects of the economies of scale.

But there are negative effects of the economies of scale also. Back in caveman days, when there were only a million people in the whole world, you could dump your wastes and throw your garbage anywhere, and it didn't matter. This was true until people started living in walled cities, where they were trapped inside with one another's garbage and sewage and (they did not realize) germs. Today, with over seven billion people in the world, we can no longer just flush our wastes into the river or throw our garbage in a pile. There are so many of us that our sewage can easily overwhelm the natural microbes that would decompose it into harmless molecules, and would certainly overwhelm nature before the water flowed downstream to another big gob of people. Don't throw that piece of paper down! You're littering! Put it in the trash can! Or better yet, recycle it, since there are so many of us using paper that we have to recycle paper rather than cut down the few remaining trees. Sure, it's annoying, but it is the inevitable result of our nonlinear population growth.

THRESHOLD VALUES

A related nonlinear concept, also ubiquitous in the natural and human world, is the *threshold value*. The term comes from the threshold of a house (originally, the beam of wood that defined the door of a threshing floor): what happens outside of the house may not matter, but once you do it inside of my house, watch out. A threshold value is a value below which something doesn't matter, a value that must be exceeded in order to make something happen. Here are four examples: how to start a fire, why clocks tick, catholes, and cows.

How to start a fire. Threshold values show up at the molecular level. Chemical reactions require an *activation energy* before they occur. Even in spontaneous reactions (reactions that release energy and therefore can happen by themselves), a little shot of energy is necessary to get them started. One example is combustion. Wood reacts with oxygen and produces heat, water, and carbon dioxide. But this will not happen if the wood just sits there exposed to air. It takes a spark to start the fire. Once the fire has started, it continues burning until the wood has been consumed. The spark is the activation energy.

Why clocks tick. Another example of a threshold value is the second hand of a battery-powered clock or watch. The electricity provides the energy to move the second hand. But the voltage must exceed a threshold value before the second hand moves at all. That is why the second hand ticks instead of just grinding along. As the battery runs down, it produces less voltage, and this could cause a non-clicking second hand and the whole clock to gradually slow down. It wouldn't take very long before such a clock would become very inaccurate. But because a threshold voltage is required to move the second hand, it will move just the right amount every time it receives sufficient voltage. When the voltage is insufficient, the second hand just stops. It might wiggle, but it stops ticking. In this way, the clock gives accurate information right up until the moment the battery voltage is below the threshold value.

Catholes. There is also a threshold value of waste that a system, such as a forest, can tolerate. Back when John Muir was hiking around in the Sierras, he could poop anywhere he wanted, and it didn't matter, because his poop was below the threshold value that the forests could process. But today there are so many hikers in the Sierras that they have exceeded the poop threshold. They have to dig catholes (this is in fact the official National Park Service term for it) to bury their wastes. In some cases, there is too much hiker poop even for catholes to accommodate, and the National Park Service has to install portable toilets, with the fleeting hope that Congress will give them enough money to clean them out once in a while (figure 7-1).

Cows. Another example of a threshold value is cows in a pasture. For any pasture, there is a certain threshold cow population that can graze without causing any damage. But once that threshold is exceeded, the result is overgrazing. (What else could it be?) The cattle eat the good grasses faster than the grasses can grow back. The thick sod of grass roots had formerly kept out noxious weeds such as spurge, milkweed, and bull nettle. But these weeds can now grow in between the cow-damaged grass clumps. Then the nonlinear processes really get into gear. Cows are not noted for their intelligence, but they are not stupid enough to eat spurge, milkweed, and bull nettle. The cows continue to deplete the grass while the weeds grow more and more. The weeds start to crowd out the grasses. And it gets worse. The pasture spirals nonlinearly out of control. Wherever overgrazing leaves bare ground, the soil can erode. The good grasses cannot grow where the soil has eroded away; instead, weeds grow in the eroded soil. The weeds have shallow roots and do not live long, so when they die, erosion resumes with a vengeance.

Figure 7-1. Every natural system has a threshold level of wastes above which it cannot process them. (Cartoon by Leslie Gregersen.)

Once again, this did not matter in the old days when cowboys could run their cattle over hundreds of square miles of prairie, or previously when bison and Native Americans roamed over that prairie. But modern ranchers have to carefully calculate their stocking rate of cattle in order to prevent pasture degradation. Some do. Many do not. Once the threshold cattle density is reached, nonlinear processes cause the pasture to rapidly degenerate.

DISCOVERY CURVES

Here is an example of a non-linear process that you can try out for yourself. It is something that scientists delightfully call a *discovery curve*.[4] If you take a walk outdoors and keep track of the number of different kinds of plants that you see, you would expect a linear process of discovery: you walk ten minutes and discover a certain number of plant species, and you walk twenty minutes and find twice as many. But it doesn't work that way. The greatest number of kinds of plants that you see during any given time interval is right at first as you look around the place you are standing. As you walk along, you will see new kinds of plants but fewer and fewer new kinds as the minutes pass.

I decided to give this a try, just before writing this paragraph. I walked along a hiking path in Tulsa, Oklahoma. The path leads through relatively disturbed areas along a riverbank and then a drainage ditch. (You'd be surprised how many species of plants and animals live in drainage ditches.) I counted the number of plant families I saw in the place where I started. (A family is a group of related plant species. There were too many species for me to record them individually.) Then I walked for fifty minutes. During each ten-minute interval, I kept track of how many *new* flowering plant families I saw that I'd not seen previously. My total was forty-four.

The number of new plant families that I saw in any given time interval started to taper off. I saw twelve families without having to go anywhere, and ten new families during my first ten minutes of walking. But in the last ten minutes, I saw only two new plant families.

If you want to try this out for yourself, it helps to know what all the plants are. You could probably do it even if you didn't know what they are, but just how to recognize which kind of plant you've not seen previously. Bird-watchers have the same experience. They see most kinds of birds right from where they are standing, but as they keep hiking, they see fewer and fewer new kinds of birds. (It's not fair to hop in the car and drive to a new place to continue your discovery curve. The curve works only in any given place. And, of course, just as bird-watchers would not include a caged parakeet on their life list of observations, you should, as I did, exclude anything that was deliberately planted.) If you like cars, you could do a discovery curve with them instead of plants or birds.

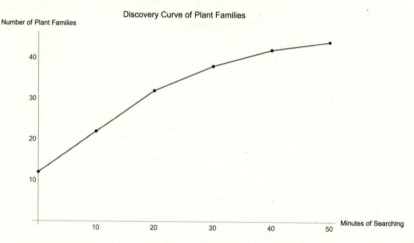

Figure 7-2. The cumulative number of plant families the author encountered on a fifty-minute walk. The number of newly encountered plant families decreases over time, resulting in a discovery curve, not a line. (Graph by Christopher Moretti.)

A discovery curve can be skewed a little bit by observer bias, about which I will talk in a later chapter. When I began my walk, I just glanced around. But by the end I was climbing around and looking wherever I could. That is, my *observer intensity* was not constant; it increased as I went along. Despite this change in observer intensity, however, my discovery list formed a curve, not a line.

THE NONLINEAR EXPLOSION OF EVOLUTION

If you look at an evolutionary timeline in any science book, the thing that most strikes your attention is that the increase in biodiversity is non-linear. Single-cell life-forms evolved about 3.5 billion years ago, and for the next three billion years it looked as if nothing was happening. If you had visited Earth during that time, you might have seen patches of green scum in the ocean but not much else. For three billion years! But a half billion years ago, multicellular life-forms exploded in abundance and diversity.[5] And the rate of this explosion has accelerated through time, except when mass extinction events hammered Earth 250 million years ago and again sixty-five million years ago. What you do *not* see is a gradual, linear increase in the number of species and their complexity.

The reason for this is that multicellular life-forms, once they began to evolve, formed a network of interactions. They provoked one another into further diversification. As soon as the first predators evolved, the prey evolved lots of new ways of defending themselves, which provoked predators to evolve new ways of eating them. As soon as the first flowers evolved, insects evolved ways of pollinating them, which provoked flowers to evolve into yet new forms to attract different pollinators.[6] Species interactions are the feedback dynamo that ignited the explosion of evolution that continues today, and will continue ever faster unless humans or a natural disaster cause another mass extinction event. Non-linear processes can affect entire planets.

THE EFFECT OF RARE EVENTS

Sometimes a rare occurrence can completely alter the course of events, making them extremely different than a linear projection of events would have been. I'm sure you can think of lots of examples from human history and current events. Let me use an example from evolutionary history.

Monkeys evolved in Africa. Many species of monkeys (the Old World monkeys) are still there, and some lineages of Old World monkeys evolved into apes. There are no apes in the New World (North and

South America). But there are New World monkeys. It would seem to make sense that, at one time, monkeys ranged freely between what is now Africa and what is now South America, back before the Atlantic Ocean had formed and when Africa and South America were a single continent. Only then did these two groups go their separate evolutionary ways. It turns out, however, that the explanation is not quite so simple. Genetic studies indicate that the divergence of the New and Old World monkeys occurred *after* the Atlantic Ocean formed. This means that African monkeys had to cross the Atlantic to get to South America and evolve into New World monkeys. Even though the Atlantic Ocean was not as wide as it is today, this would still have been a spectacularly unlikely event. Maybe a floating mat of vegetation from Africa washed up on the shore of South America, with some monkeys riding on it. Though rare, such events have been observed.[7]

THE NONLINEAR DEATH OF MARS

Earth's sister planet Mars had a nonlinear death.

Ever wonder why Earth is alive but Mars is dead? Maybe not. Monotheistic religions tell us God created life on Earth and not on Mars, and that's that. But scientists are not satisfied with explanations that end with "that's that." (Besides, the Old Testament, on which three monotheistic religions are based, does not say that God didn't create life on Mars.)

Of course, maybe Mars didn't used to be dead. There is a great deal of evidence that Mars used to have whole oceans of liquid water. And it may have had bacteria-like organisms.[8] Early in the history of the solar system, Earth and Mars were not too different from one another. Then something happened. Something nonlinear.

Mars is, of course, farther from the sun than is Earth. It is, in fact, one and a half times as far from the sun as Earth is. This would mean that sunlight intensity (which changes nonlinearly with the square of distance) is two and a quarter times dimmer than on Earth. Not surprisingly, then, Mars is colder than Earth. But this may not always have been the case. Mars may have had, soon after it formed, a thick atmosphere that consisted mostly of carbon dioxide, giving it a prodigious greenhouse effect, perhaps enough to make it moderately warm four billion years ago. Today, the atmosphere of Mars is very thin, but what little atmosphere it has still consists mostly of carbon dioxide, allowing it to not be as cold as it would otherwise be. At the equator in the Martian summer, temperatures can get up to 70° F (20° C).[9]

The radius of Mars is almost exactly half that of Earth, and its gravity is about 38 percent as strong as that of Earth. One might therefore expect its atmosphere to be about 38 percent as dense as that of Earth. This is not, however, the case. The Martian atmosphere is less than 1 percent as dense as ours. (If you hear that the wind on Mars blows at six hundred miles an hour, do not assume that it would feel like a six-hundred-mile-per-hour wind on Earth.) What happened to the missing Martian atmosphere?

Mars is not much smaller than Earth, but it is small enough that, apparently, it was below the threshold value for retaining its mantle of molten lava under the surface. Its mantle cooled off enough that it no longer had any lava circulation. In contrast, the lava in Earth's mantle still circulates. As it does so, it creates a magnetic field that deflects most of the harmful radiation coming from the sun (called the "solar wind"). Once the inside of Mars cooled off (which would have had very little effect on its surface temperature that, like Earth's, is determined by sunlight intensity), it lost its magnetic field and was bombarded by solar wind that took away much of its atmosphere, including nearly all of the water. There's still some water underneath the Martian surface, but the surface of Mars is now dry.

That is apparently how Mars died. Mars is 50 percent smaller than Earth and has only 38 percent as much gravitation, but this relatively slight difference was enough to allow it to die and Earth to live.

THE NONLINEAR EXTINCTION OF HUMANS

Will *Homo sapiens* ever become extinct? Yes, of course, at least when the sun expands into a red giant and consumes all of the inner planets, but this will not happen for several billion more years. (Unless some humans get in a space ship and go to another planet, which will take several human generations.) But could it happen prematurely? Maybe in just the next few centuries?

At first, premature human extinction might seem impossible. Humans have enormous creativity and have figured out how to live in practically every habitat on Earth.

A quick perusal of websites that address human extinction reveals that nearly all of them discuss factors, such as nuclear holocaust or an asteroid, that would directly kill all humans. But they are all, I believe, overlooking the fact that most humans depend upon a global society. A few scattered tribes of humans might survive for a while after the collapse of civilization, but human extinction would quickly follow. The major human adaptation is culture. Perhaps there are no humans who can survive without the cultural network of other humans—collective knowledge, mutual aid, etc.[10] I know I couldn't. Could you? Self-styled "survivalists" would die as soon as they ran out of ammunition. Once the human population of the world becomes small enough, it will rapidly collapse because the surviving tribes will have insufficient cultural knowledge to make a living even on an Earth that is not deadly or devoid of food.

The Neanderthals, a species of humans that existed before *Homo sapiens*, became extinct in this fashion. Modern humans drove the Neanderthals out of the best land by about thirty thousand years ago. The last Neanderthals clung to existence in hostile environments. When the last Neanderthals died, there was still food and other resources on Earth, but their population sizes had become too small to maintain their culture, which included techniques of toolmaking and hunting.

Human extinction will not occur in a linear fashion. There is a threshold below which humans will not be able to maintain a culture of survival. It has happened before. And nobody knows what that threshold may be.

DO IT YOURSELF!

1. Do your own discovery curve!
2. Think of some process in the world around you that you have always assumed was linear but turns out to be exponential or has a threshold value.

CHAPTER 8

IT'S NOT ALL BLACK AND WHITE

On the moon, it's all black and white. Night is night, day is day, and there is nothing in between. On Earth, however, there are twilight shades of gray because the atmosphere scatters sunlight. And that is just the first of almost countless ways in which Earth has, and science studies, diversity.

But the recognition of diversity does not always come easily to the human mind. The human mind has a binary bias—we see things as black and white—and science must resist this bias, just as it resists many others. Even when the human mind does not look at the world in a binary fashion, it looks at the world in a categorical fashion: we like to put everything into categories, even though reality often consists of a continuum rather than discrete categories. Binary thinking is a two-category subset of categorical thinking. Want some examples of categorical vs. continuous? That's what this whole chapter is about.

Maybe our human penchant for categorical thinking comes from the fact that we are bilaterally symmetrical creatures. We have right and left hands, feet, and lots of other things. We also have a top and a bottom, and a front and a back. In contrast, a jellyfish is radially symmetrical: it has a front and a back, but otherwise its symmetry radiates out from a central point. If a jellyfish could think, it might see the world as a whole range of possibilities rather than as this vs. that. Or our binary thinking may have evolved due to the fact that our animal and human ancestors had to decide, from one moment to the next, whether to act or to not act—to flee from the tiger or to not flee, to eat a certain kind of food or not eat it. Life-and-death decisions are often binary, even for jellyfishes.

On the one hand, humans tend to see the world as white or black, this or that, left or right, here or there, up or down, us or them. On the other hand (to continue my bilateral metaphor), we recognize that there is a lot of diversity that cannot be contained within categorical thinking. (There are two kinds of people: those who put things into two categories, and those who don't.) We humans constantly struggle to reconcile categorical vs. continuous thinking. Assuming (that is) that we can categorize all thinking as either categorical or continuous.

Being binary, moreover, appeals to our sense of fairness. Reporters have a strong bias to "report both sides" of a story even if there are more than two sides and even if one of the two (or many) sides is patently ridiculous.

DIVERSITY IN THE MICROWORLD

In the physical, chemical, biological, and human world studied by scientists, few things are categorical. One of the few binary things is the electric charge of atomic particles. Electrons have a negative charge, while protons have a positive charge. But even charge is an emergent property of the quarks that constitute the electron and proton particles. Everything else seems to exist in the form of numerous possibilities, whether categorical or continuous.

Take, for example, atoms. The word "atom" means *that which cannot be divided*. Atoms can be split, but when this occurs they lose their identity, so "they" cannot be split and still remain themselves. We might think that all carbon atoms are alike, all part of one category. But this is not true. They all have six protons in their nuclei. Most of them also have six neutrons (making them ^{12}C or carbon-12). But a relatively small number of carbon atoms have an extra neutron, which makes them heavier (^{13}C, or carbon-13). And an even smaller number of carbon atoms have two extra neutrons, which destabilizes the nucleus and makes them radioactive (^{14}C, or carbon-14). Different isotopes of an element have the same number of protons but different numbers of neutrons. Similarly, a pure iron atom has twenty-six protons and twenty-six electrons. But many iron atoms have lost some of their electrons. Ferrous ions have lost two electrons, and ferric ions have lost three. This gives them different electrical charges. Different ions of an element have the same number of protons but different numbers of electrons. Each kind of element, then, consists of different types.

Within each of these categories, we might think, the atoms are all alike. But even this is not entirely true, because atoms never exist in isolation. Consider, for example, the two hydrogen atoms and the oxygen atom that make up each molecule of water (H_2O). The atoms within a molecule share their electrons freely. But not entirely freely. Oxygen atoms are notoriously hungry for electrons. Within an H_2O molecule, the freely moving electrons spend more of their time with the oxygen atoms and less with the famously wimpy hydrogen atoms. Water molecules have a neutral charge—each one has a total of eighteen protons and eighteen electrons—but it has three poles: two positive hydrogen poles and one negative oxygen pole. The positive poles of one water molecule are attracted to the negative poles of other water molecules, making water molecules stick together just a little. This stickiness of water molecules is an essential property that causes ice to float, liquid water to hold a lot of heat before it finally boils, transpiration to pull water through plants, and many other things without which life would not be possible. These "hydrogen bonds" also hold the strands of DNA together just strongly enough to preserve molecular integrity but just loosely enough to allow the strands to disassemble and reassemble. (DNA is the molecule that stores genetic information in a cell. In order for this information to be used by the cell or passed on to the next generation, the strands have to be able to separate from one another, revealing their hidden information.) The characteristics of an atom, therefore, depend on which other atom or atoms to which it may be bonded.

One atom or molecule can change another, even without bonding with it. The electrons of one molecule can (to use a metaphor) scare away the electrons of another molecule next to it, resulting in a charge difference that can allow the molecules to attract one another. Such "Van der Waals forces" allow geckos to run up walls without actually sticking to them. In 2014, DARPA (the Defense Advanced Research Projects Administration) announced the development of a gecko suit that uses these forces to allow soldiers to scale walls, not as easily as geckos but more easily than any other soldiers.[1] Within each ionic or isotopic category of atom, there is a whole continuum of possible characteristics, depending on other atoms with which it may be bonded or even those it may happen to be close to.

Many molecules can exist in more than one conformation, for example as different mirror images of one another. (There are only two possible mirror images for such molecules; this might be one of the few binary properties in the universe other than charge.) This makes a great deal of difference. "Left-handed" amino acids make up the proteins that keep us alive; "right-handed" proteins are often poisonous. A mixture of left- and right-handed amino acids results in an unstable protein. Natural selection would therefore have eliminated any proteins that were a mixture of left- and right-handed forms. Life on Earth just happened to get started with left-handed amino acids, leaving the right-handed ones to fill the villain's role. Maybe on Mars, before Martian life's nonlinear demise, the proteins consisted of right-handed amino acids, if it had proteins at all.

And it gets even more diverse. Consider a bunch of molecules, all of the same kind, with no differences of charge or handedness. This collection of molecules has a certain temperature. Temperature results from the energy of movement, or kinetic energy, of the molecules. But no two molecules have exactly the same kinetic energy. Each one has its own unique energy state, some moving more, some moving less. You can think of temperature as the average kinetic energy of the molecules, although it is not strictly speaking a mathematical average. When water boils, the average kinetic energy is enough to cause the molecules to stop cohering to one another in a liquid and to fly off in free gaseous movement. But even before water boils, many of the water molecules have enough energy to evaporate. Even ice has a few water molecules that can launch into a gaseous state—a process called, in one of those rare instances of beautiful scientific terminology, sublimation. So even the water molecules in a glass of water have diversity. And it is continuous diversity. The molecules are not in kinetic energy categories.

The number of possible kinds of molecules that can exist is theoretically infinite. In the real world it is not infinite, but it certainly exceeds the ability of the human mind to comprehend it. Or, at least, of my mind. I got a C in organic chemistry in 1976, and it's been downhill from there.

So it's diversity, diversity, diversity, often continuous instead of categorical—and we haven't even gotten beyond molecules yet.

Molecules are also diverse in what they do. Here is just one example. A nerve impulse occurs when protein gateways in a nerve cell membrane open up and allow sodium ions, previously kept out, to rush into the nerve cell. But what actually happens is that *some* of the protein gateways open and some do not; only when these proteins reach some kind of a consensus with one another, or a quorum, does enough sodium flow into the cell to stimulate a nerve impulse. The proteins do not all act exactly the same way all the time.

DIVERSITY IN THE MACRO WORLD

Nor are any two cells, even of the same type, exactly alike. Take, for example, red blood cells. There's not much to a red blood cell; it's pretty much hemoglobin inside a membrane. The average red blood cell lives 120 days. But very few, if any, red blood cells actually live 120 days. One hundred twenty days is an average. Some of them live 100 days, some 140; some live a few hours, while a few may live for a year. There is a whole continuous range of lifetimes found among red blood cells. Or any other cell.

Incidentally, the same is true of human life spans. The Bible says that humans live seventy years. There are some fundamentalists who take this literally and think that they will drop dead at age seventy. I'm not joking. But each group of people, whether a family, a city, a country, or a race, has its own average life span. The recent increase in average life expectancy is not so much due to a prolongation of maximum life span—there have always been very old people, even in ancient times— but a decrease in infant and childhood mortality. This raises the population average age at death.

Even whether an organism is dead or alive is not entirely a binary state. Being dead or alive is not, of course, an entirely continuous state either. There is a point at which a body, due to the accumulated ravages of old age or of injury, shuts itself down. The "time of death" on a death certificate is not a mere average of what all the atoms, molecules, cells, and organs are doing or not doing. The "moment" of death is not an illusion. But neither is it entirely binary. People who are "brain-dead" do not really have dead brains, but the only part of their brains that still functions is the brain stem, which maintains breathing, heartbeat, and digestion. And the "dead" part of the brain is metabolically alive; the nerve cells in the cerebellum and cerebrum continue to metabolize—but they do not do much else. Is a brain-dead person alive? To a certain extent, death is now at the mercy of technology. Lungs can be stimulated to breathe, and hearts to beat, by medical intervention. At the "moment of death," many metabolic processes continue on the cellular level for at least a few minutes. That's why the blood of a dead person is brownish instead of red: the cells continued to take oxygen from the blood even after the person stopped breathing. Unless the person died from cyanide poisoning. Cyanide prevents cells from using oxygen. A cyanide-poisoning victim therefore has bright red blood hours after death.[2]

No two organisms within a species are exactly alike. Sometimes two individuals can be "clones" of one another, which means that their genes are exactly the same. (Genes, the basis of *genetics*, are information encoded in DNA and can be passed down by sexual reproduction from one generation to another.) The most familiar example of clones is identical twins. But genetic identity does not guarantee that the "clones" will be identical in their appearance or behavior or thinking. Different individuals, even with identical genes, have different experiences both before and after birth. Malnutrition or injury can cause bodily changes. Different experiences can cause psychological differences. Some physical experiences (such as malnutrition) can cause differences in the activation status of the genes—that is, the way that the genes are used. Therefore two individuals with the same genes can use these genes differently. And the activation status itself can be passed on from one generation to the next—this is the basis of the astonishing new science of *epigenetics*. Given all these possibilities for bodily and psychological diversity, it is astonishing how much alike twins can be, even when raised in total isolation from, and ignorance of, one another. Sometimes identical twins can choose to diverge from one another, perhaps to consciously assert their individuality; sometimes they can choose to imitate one another, perhaps to generate a feeling of companionship, or perhaps just to see the looks on our faces when we keep getting them mixed up.

Our binary categorical bias encourages us humans to distinguish "us" from "them." In so doing, we tend to homogenize both. We dehumanize our enemies as being all alike, and ourselves as well, assuming that we are, or at least should be, all alike. If our enemies are not all alike, if we realize that some individuals are good and some are bad, then we might feel that it is unfair to drop bombs indiscriminately on them. And if *we* aren't all alike, the inconvenient minority can be classified, perhaps, as traitors. Take, for example, the United States before the Civil War. Not all Northerners were abolitionists; in fact, some profited from slavery, even if they did not themselves own slaves. Northerners were required by law to capture fugitive slaves and return them to their masters. And not all Southerners were cruel torturers of slaves. But the conflict between North and South tended to dampen down the expression of the minority viewpoint in both. Slavery sympathizers in the North and abolitionists in the South kept their mouths shut. Social adversity tends to propagate uniformity. We see this in our current political situation, in which moderate politicians who cooperate with members of the other party are becoming ever less common.

Perhaps social homogenization within categories is most obvious in the matter of racial identity. An intelligent visitor from another planet might be truly puzzled by the concept of human races. While races are not entirely an artificial construct, racial classification does tend to hide a great deal of diversity. But racial differences are not all black and white. Just how many races are there, anyway? For example, the Bantu people, who are black and whose ancestors originated in western Africa, have relatively large lips and flat noses. But the San people, who are also black and whose ancestors originated in southern Africa, have very different facial features from the Bantu, and skin that is sometimes more yellow than black. Ethiopian blacks have longer faces, smaller lips, and more pointed noses. And from there it gets complicated. Apartheid South Africa struggled with such racial definitions and finally had to classify people whose ancestors came from China as honorary whites.[3] This was a laughable (if it were not so serious) example of a desperate attempt to impose categories on continuity.

Then there is the matter of racial mixture. Many people, including myself, are of mixed ancestry. It is astonishing that the US Census Bureau did not until recently allow an individual to identify multiple racial identity. As Brian Sykes explains, some people who consider themselves pure black or pure white may discover some DNA from a different race hiding in their chromosomes.[4] Here in Oklahoma we have black Cherokees.

And from there, things get even more complicated.

There are genetic and even racial differences in every species of microbe, plant, and animal on Earth. We do not even know how many species there are on Earth. When I was a kid I used to think that there was a big filing cabinet somewhere where all the species were documented on sheets of paper. I suspected that the cabinet was at the Smithsonian Institution and that the deep-voiced Loren Eiseley (who hosted a Saturday morning show about the Smithsonian) had the key. But there is no such repository of species information. Websites such as the Encyclopedia of Life have a webpage for each known species, but these websites are and will forever remain works in progress.[5] Biologists estimate that there may be ten million species of organisms, depending on how you define species. Biologists attempt to define species as organisms that crossbreed with one another in the natural world. But then some species hybridize with one another. Species are true, but not entirely discrete, categories. Biodiversity is a lumpy gravy, the lumps being species.

Likewise, it is impossible to categorize ecological relationships among species. One morning, as most mornings in Oklahoma, I saw vultures flying overhead. Vultures are scavengers, and that's that. Or is it? Vultures, like all scavengers, specialize on dead meat. Dead meat does not fight back or run away. But "scavenger" is not a vulture's identity, just its specialization. Any other kind of meat that cannot fight back or run away, even if it is still alive, may serve as well for a vulture's repast. Oklahoma ranchers know that when a cow is ready to give birth, they had better either have the cow in a barn or be out in the pasture with the cow because vultures will watch for the moment the little calf is born. At that moment, the vultures move in and start pecking away at the tender spots on the newborn calf, especially the eyes and the anus. "Scavenger" is a category, but there is some spillover between this category and the "carnivore" category.

Similarly, cows and deer are clearly in the "herbivore" category. Cows walk around and moo and eat grass all day. So adapted are they to their herbivore category that they have specialized bacteria in their stomachs that assist them in digesting grass. Grazing (eating grass) and browsing (eating plants other than grass) are different categories of herbivory, and herbivores may specialize on one or the other. Generally speaking, cattle graze and deer browse. But they can sometimes violate the herbivore category. Cattle have been known to eat the heads off birds caught in mist nets, and people in rural Oklahoma who raise chickens know to use two layers of chicken wire to protect the chicks not just from raccoons but also from deer, who would otherwise reach their muzzles into the coop and feast on hatchlings. Squirrels eat nuts. But if a squirrel gets run over by a car, another squirrel may run out to it, not to say goodbye to its erstwhile friend but to sneak in a few bites of meat before the next car comes along.

Scavenger and herbivore are perfectly valid categories. Each species is a valid category. But scientists recognize that categories are almost always imperfect, as when herbivores eat meat or individuals within two different species hybridize with one another. Race is a useful category for descriptive purposes, although most people now recognize that every person is a unique genetic mixture. Some scientists think that races will eventually blend together, as they are now doing in Hawaii and Brazil;[6] others think that people will continue to prefer mating with members who have the same external racial characteristics that they have, thus maintaining racial distinctions. But there is no such thing as categorical purity—of scavengers, herbivores, or especially of human races.

In exploring the physical world, science embraces all these levels of diversity. For some purposes, order must be imposed on this diversity. The fact that we cannot be absolutely sure if our classification of things is correct does not prevent us from studying them anyway. We continue to study the psychology and health of "humans" even though no two humans are exactly alike. The fact that not all membrane proteins are in complete agreement with one another does not stop nerve impulses from happening. Scientists measure temperature, even though each molecule has its own level of kinetic energy. We scientists look for patterns and test hypotheses as best we can, not letting the diversity overwhelm us. Fortunately, most of us enjoy diversity, whether the natural diversity of seeing lots of different tree species, or the cultural diversity of ethnic cuisines.

RELIGION AND CATEGORICAL THINKING

Categorical vs. continuous thinking is also reflected in our religions. On the one hand, monotheistic religions are binary. There is one God and he is our God; the gods of other religions do not exist, or are demons. And everyone who has ever lived is either going to go to heaven (or paradise) or to hell. (The Catholic Church used to recognize a third zone, the *limbo*, into which innocent dead babies went. Recently, however, it discarded this concept. Dead babies now go to heaven.) Old-time religion classifies people into "sheep" (saved people) and "goats" (damned people). No matter how much diversity there may be, in the end it does not matter: even though sheep and goats are similar enough that they can be hybridized (forming "geeps"—I'm not joking), God the Great Judge will separate them and send the sheep to safely graze the pastures of heaven forever and send the goats down to chomp forever on the trash piles of hell. On the other hand (I just can't seem to get away from this binary thinking!) polytheistic religions may see an essentially infinite number of possible fates after reincarnation. The biological world is immensely diverse, yet Genesis 1 imposes strict categories on this diversity: plants created on day 3; fishes and birds on day 5; and humans, cattle, and "creeping things" on day 6. (It is interesting that, in this biblical chapter, humans do not get a day or category to themselves—they are thrown in with the creeping things.) Later in the Torah, animals are all classified in a binary fashion: clean vs. unclean.

There is another way in which religion creates false binary categories, and this time it is an opinion shared by many scientists. Many people assume that science and religion must be enemies. Centuries ago, Galileo's explanation that the earth goes around the sun was considered an attack on religion. Today, few people believe this (there still are some) but continue to oppose evolutionary science by assuming it is an attack on God.[7] It is true that many religious people try to use their faith to dispute scientific facts. But many scientists are personally religious. One of them was Galileo himself. We should remember what Galileo said: religion "tells us how to go to heaven, not how the heavens go."

DO IT YOURSELF!

1. Think of something that you have always considered uniform and look closely at its diversity, whether it is roses in your garden or salt crystals under a microscope.
2. Want to read a novel about the desperate attempt to force every individual into a racial category? Try Mark Twain's *Pudd'nhead Wilson*.

CHAPTER 9

CAUSE AND EFFECT

In Japan, some people practice *shinrin-yoku*, which is bathing in the scent of a forest. This practice has been shown to reduce blood pressure and the amount of the stress hormone cortisol in saliva.[1] Some researchers attribute the stress reduction to volatile chemicals, such as monoterpenes, released by the trees.[2]

This brings us face-to-face with one of the most important concepts in all of science: correlation is not causation. Two variables can be correlated, such as high blood pressure with the absence of monoterpenes and low blood pressure with the presence of them. This does not mean that the monoterpenes caused the blood pressure to be lower. Our brains are biased to interpret correlation as causation. We do it literally without thinking.

You see, what was probably happening out in the forest was that other factors were causing people to relax. When the people were out in the forest, monoterpenes were not the only thing they were experiencing. In every other way, they were relaxed. No hectic schedule. No noise. No other people. Shade. Susurrus (look that up). A mental awareness of the happiness of being in nature, something Edward O. Wilson, perhaps the most famous scientist in the world, first called *biophilia*, the love of life.[3] And bias, also: the people *expected* to feel relaxed in the forest. Oh, and monoterpenes too. The investigators were aware of this problem. They administered monoterpenes to laboratory mice under controlled conditions and found physiological effects similar to those of humans, compared to controls.

Statistical analysis, by itself, cannot solve this problem. Statistical methods can calculate correlation coefficients and tell you whether they are or are not significant, but that is all. So when the *New England Journal of Medicine* published a paper in 2012, claiming that eating chocolate could make you smarter (oh, don't we wish!), the authors had seen a spurious correlation.[4] Smarter people do lots of things differently from people who are less smart; apparently, one of them involves eating chocolate.

MULTIPLE AND HIERARCHICAL CAUSATION

The people relaxed in the forest for multiple reasons, one of which may have been volatile chemicals. This is an example of *multiple causation*.

There is another way in which an effect can have more than one cause. The causes might cause one another in a hierarchical series. This is *hierarchical causation*. Suppose someone took a gun and shot you. Naturally, you would say that the man shot you. But you could also say that the gun shot you, or that the bullet shot you, or that the laws of chemistry and physics (which explosively imparted momentum to the bullet) shot you. You could even say that God shot you, if indeed God is in charge of natural laws. A hierarchy of causes. It sort of sounds like "This Is the House That Jack Built." You know, the nursery rhyme that goes something like, "This is the dog that chased the cat that killed the rat that ate the malt that lay in the house that Jack built."[5] While this may seem whimsical, I remind you that a multibillion-dollar industry is built on the perhaps deliberate misrepresentation of hierarchical causation. The American gun lobby defends the millions of guns in the United States (estimates vary from two hundred to three hundred million[6]) by saying, "Guns don't kill people; people kill people." Of course, nobody claims that guns come out of their cabinets and kill people. People use guns to kill people.

Reality consists of a complex mesh of hierarchical and multiple causations. Consider a hiker encountering a bear on a trail. What happens next? Will the bear attack? Will it turn and leave? Will the attack be fatal or just result in a mild injury? There are lots of factors at work. Some of the factors are related to the person. What was the person doing at the time of the encounter? Were other humans present? How does the person or how do the persons act toward the bear? Does the person have something to scare the bear away with? Then there are factors related to the bear. What the bear does next depends on its species (black bears might be less dangerous than grizzlies), its gender, its mood, whether it is hungry or not, its individual behavior patterns (there are apparently a few psychotic bears out there), how many other bears there are in the vicinity, whether the bear has had good or bad or no prior experiences with humans, whether the bear is a dominant or a subordinate member of its own society, whether the bear sees or smells the person or people, etc. If it is a mother bear, the reaction will be different if she has cubs than when she does not. There are environmental factors too: the reaction may depend on the habitat, time of year, time of day, etc. Of course, you would not have time to think about any of this. Some people say you should make yourself look bigger if you encounter a bear. But this might only make the bear think it has a bigger meal waiting for it.

We find multiple causation almost everywhere we look. For example, one of the effects of global warming, and its resulting warmer winters in the temperate zones, is that many individuals in migratory bird species are now staying home because the winter is shorter and warmer.[7] But that is not the only reason. In the old days, birds had to migrate to find food in the winter, but these days there are thousands of people who have bird feeders. Maybe some of the birds stopped migrating not because of warmer winters but because of bird feeders. Bird feeders are not likely to have a major impact on bird migration, since the food they provide is considerably less than what their whole populations require. But bird feeders are one of the multiple reasons that some birds have stopped migrating.

CAUSE vs. EFFECT

Finally, it is sometimes difficult to distinguish which factor is the cause and which is the effect. And it can make all the difference in the world which is which. It is well known that countries that have a high population growth rate also have a lot of poverty. (That does not mean that they are poor countries. There could be a small upper class of rich people and a lot of poor people, resulting in an average level of wealth with which most people would be comfortable.) The natural assumption is that the people are poor because they have too many kids. If this is the case, then the future of humankind is pretty bleak: if you give food and medicine to the poor, they will just have more kids, and you will end up with the same level of poverty, only more people will be poor. Attempts to relieve misery only end up creating more misery. This is the "utterly dismal theorem" of economist Kenneth Boulding.[8]

But what if we reverse the causation and say that poverty causes high birth rates? This sounds ridiculous at first, but consider a family that lives in a country without any economic or health security. If such a family has only two kids, they might both die. In a family with more kids, there is a chance that one of the kids will get a good job and provide more resources to the family. In case this still seems incredible, consider natural selection, about which there is a chapter later in this book. Natural selection rewards individuals, not groups. An overpopulated country might be miserable, but natural selection favors individuals (and families) that win in the game of competition. Now, if this is the case, then providing food and medicine will actually cause the birth rate to decline, since parents may choose to have fewer kids. (This also presupposes a society in which such a choice is available, e.g., through contraception.)

The fact is poverty and the high birth rate are correlated, but we cannot tell which causes which simply by looking at the correlation. It requires experimental testing. The experiment has unintentionally been done. Countries such as the United States have been unable to stand aside and let people in India and most countries in Africa suffer. So they have provided food and medicine, even at the risk of making the population explosion worse. But guess what? In nearly every country in the world, over the last few decades, the birth rate has *declined*.[9] Therefore, poverty causes the population explosion, and by working against poverty, you are also helping to solve the population problem. This is good news. Cling to it, my friends, for we need all the good news we can get. World poverty is, in some places, declining; starvation is now rare in India. World population growth is already leveling off, although it may not do so until after the human population has stressed the planet so much that its ecosystems fail.

Sometimes the cause-and-effect arrow goes both ways. When this happens, watch out. We call it "the vicious circle." An important example involves global warming. Higher temperatures are causing polar ice to melt. But when the ice melts, the polar land and oceans absorb more and reflect less light, which causes the polar environment to become warmer. Global warming causes ice to melt, which causes global warming, which causes more ice to melt, which . . . round and round we go, and where we stop, nobody knows (figure 9-1). I used to worry about the population explosion. Now I worry about this. Of course, population growth makes global warming worse. Pass the wine.

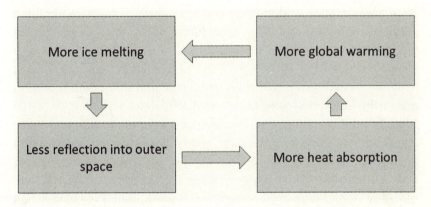

Figure 9-1. Global warming is an example of the vicious circle in which each effect becomes a cause that reinforces the effect.

An article appeared in the journal *Science* in 2014 that was a particularly interesting example of back-and-forth networks of causation.[10] It involves the origin of agriculture.

First, some background on the origin of agriculture. Most people assume that agriculture was something that humans invented. In particular, some brilliant man invented it one afternoon in the ancient Near East. He said to himself, "We go around gathering wild seeds, which means we have to keep moving. But if we planted the seeds, and they grew up into the next generation of plants, then we could just stay home and eat them." And thus not only was agriculture born but also village life. This story has been called the "brilliant man" theory of the origin of agriculture. Then some other people pointed out that it was actually the women who gathered most of the grains, fruits, tubers, and nuts. So women must have invented agriculture. This would be the "brilliant woman" theory.

But the origin of agriculture had to be a more gradual process.[11] Consider the origin of cultivated wheat, the earliest Eurasian crop. If someone had tried to plant wild seeds, they would not have grown. Wild grains have germination inhibitors in them. Also, in wild grains, the seeds fall off of the stem; you have to gather them before they fall into the dirt. Moreover, wild grains tend to be small, in contrast to plump grains of modern wheat. All of these things would be very inconvenient for agriculture. So if a brilliant man or woman had tried to invent agriculture, using wild grains, it would have been a failure. But instead what apparently happened is that people who gathered wild grains tended to gather the grains that were large and that had not yet fallen off of the stem. That is, the grains that they gathered were not a random sample of wild grains; they were the large grains that clung to the stems and grew when sown. They took these seeds with them when they moved to a new location. With each generation, they gathered seeds that were more like our modern crop plants. After a few hundred years, they had modified wild grains that were now ready for a successful experiment with true agriculture.

However it got started, humans *caused* the change of wild plants into crop plants. But then, guess what? The plants *caused* the humans to change. Animals in general have the ability to digest starch, using an enzyme called amylase. Cultivated grains have more starch than wild fruits and nuts. As a result, people who lived off of agriculture evolved the ability to produce more amylase than people who continued to live off of wild foods. Today, the Japanese people who eat a lot of rice have more copies of the amylase gene than the otherwise genetically similar people of Siberia, who eat more wild fruits and nuts.[12] Humans caused plants to evolve, and plants caused humans to evolve.

That's just the background. Now back to the new conclusions of the 2014 article. Plants were not finished with their effects on human evolution. The crops themselves exercised an important influence on human *social* evolution. Wheat is a crop that a farmer can raise by himself. But rice is a crop that, while a lone individual can raise it, is much more productive when a whole village works together to flood the fields and transplant the rice plants. Therefore, in wheat-growing regions of China, people are more individualistic in their thoughts and actions, as measured by standard psychological tests, compared to the more community-minded people in the rice-growing regions. This even showed up in the form of higher divorce rates in wheat-growing regions than in rice-growing regions. Does it take a village to raise a child? Absolutely, if your food system is based on rice, but if your agriculture is based on wheat, maybe not so much.

The arrow of agricultural causation has two heads: people modified plants, and plants modified people, both physiologically and sociologically. Cause and effect is often considered to be one of the most basic concepts of science, but as we have seen, in many cases two—or more—processes can cause one another.

A FAMOUS EXAMPLE
FROM A FAMOUS SCIENTIST

One of the most famous scientists of all time, Linus Pauling, got correlation and causation mixed up. He mistakenly concluded that vitamin C prevents colds.[13]

There is a clear correlation: people who take vitamin C supplements have fewer colds than people who do not, on the average. But people who take vitamin C supplements have a lot of other healthy habits than just taking vitamin C. They take other supplements. In general, they tend to be more watchful of their health. They get exercise and watch their diet. They are careful about washing their hands after being out in public. Their lower incidence of colds may be due to any one or combination of these other factors. That is, vitamin C may be just one of many causes of their lower incidence of colds—or it may not be a significant contributing factor at all. Pauling mistook correlation for causation. Although he cited several studies that he believed were properly designed,[14] most studies have failed to confirm that vitamin C, in itself, cures colds.[15]

As we have seen in this chapter, it is not as easy to recognize cause and effect in the natural and human world as we might presume. Scientists spend a lot of time and effort trying to avoid reaching conclusions based on spurious correlations.

DO IT YOURSELF!

Think of something that you have always considered to have a simple and straightforward cause and think of other factors that may also be causes.

CHAPTER 10

IS BARTHOLO-MEOW INTELLIGENT?

A child naturally wants to know what causes things to happen. This is the psychological kernel upon which all human knowledge, including science, is based. But the kinds of causation that make sense to a child, and upon which the entire belief systems of many primitive societies are based, and which come so easily to all of us are quite different from the scientific understanding of the world.

SEEING INTELLIGENCE EVERYWHERE

We see people do things all the time. So, it is not too far of a stretch to assume that when we see something happening, there must be a person causing it, even if this person is unseen. When ancient people felt the wind blow, they thought someone must be blowing it with his or her breath. In ancient Hebrew and in Greek, both of them languages in which the Bible was written, the words for wind, breath, and spirit were the same (*ruach* in Hebrew, *pneuma* in Greek). What is the Spirit of God like? Jesus said that you can hear the wind in the trees, and that is what the Spirit of God is like. Mythologies of people everywhere in the world have attributed natural phenomena to gods: lightning, thunder, rain, the growth of spring wildflowers, and so on. In many cases, the actions of these gods were capricious, just like the actions of humans.

And this is a natural human bias against which scientists must guard: the bias of *agency*. We tend to think that intelligent agents cause things to happen. Western religions have progressively let go of the belief that God directly causes every natural process, such as the blowing of the wind. But this bias haunts the back of our minds and comes out, at least, in our figures of speech. A typical winter weather report says that temperatures "are struggling to get up out of the teens."

Over time, people became dissatisfied with supernatural agency. Many of the things that occurred in the world were not, in fact, capricious. In Greek mythology, the sun god, whom the Greeks called Hyperion, rode his chariot across the sky. Hyperion always rode his chariot over exactly the same route year after year after year. Didn't he ever get bored? He was acting not so much like a god as like a complex sort of waterwheel, always grinding in the same way. Even in the earliest civilizations there were astronomers watching the sky, and they knew you could plan your agricultural calendar around the movement of the stars.

Except when you couldn't. Some years are colder than others. But even then, there was a regularity that could be trusted for your agricultural calendar. You can watch the process of spring budburst in trees, and it will give you a fairly reliable indicator of when warm weather has come to stay that particular year. In the Bible, Jesus said that when the fig buds have burst, you know that winter is, at last, over, even (we can infer) if that time is not the same calendar date every year. Having actually studied spring budburst, including the buds of mulberries (which are close relatives of figs), I can confirm that Jesus's statement is correct. In my data sets, frosts have occurred after the elm buds have opened but never after the mulberry buds have opened. Jesus also said that if the sky is red in the morning, you know a storm is coming. It is not the caprice of a storm god. In saying these things, Jesus was not showing supernatural intellect but just appealing to things everybody around him already knew.

Mathematicians and astronomers began to recognize the utter predictability of many recurring astronomical events, even those like the transit of Venus that occur very seldom. But this did not mean that people stopped attributing things to gods. They just began to think that God or the gods enabled events that occur without having to be individually caused. And this religious view remains with us today. Most religious people believe God has commanded the law of gravity, rather than commanding every star and leaf to fall.

But many religious people are not willing to extend this understanding back into the past. They believe that God had to make everything, as it now is, in the past, and then set it running, like a clockmaker who makes a clock, winds it up, and then lets it run. Creationists believe natural law explains what happens today but cannot explain the origins of things. Many are so consumed by this viewpoint that they will look right at the evidence for Earth being billions of years old and for organisms having evolutionary ancestors, and literally be unable to see it.

I do not mean to ridicule such people. They are just exhibiting the bias of agency, as humans have always done. And scientists can fall into this trap as well. The most common way in which this occurs in modern science is that we assume that animals—and even plants—are intelligent when, in fact, this may not be the case.[1]

INTELLIGENT PLANTS?

Intelligence of many organisms may be an illusion, even a delusion, caused by the fact that our species adjusts to the world through intelligence. But other kinds of organisms adjust to the world in different ways, ways that we would not necessarily call intelligence. I would define intelligence as a mental process that takes data from the world and compares it with stored information in order to reach a conclusion about what to do, *and* has some awareness of the process.

Take, for example, trees. Trees seem to know a lot. Their stems always grow up, and their roots always grow down. How do they know how to do this? Well, of course, they do not actually know. They are just responding to their environment. Little starch grains, settling inside their cells in response to gravity, "tell" them which way is down, just as the little mineral-studded filaments in the semicircular canals of our ears "tell" us which way is down. Plants respond to this information by producing hormones. We respond to this information with our brains.

Plants do not have brains and do not have intelligence. Some scientists and science writers say that plants are intelligent, but only at the expense of enlarging the definition of intelligence to mean any kind of beneficial response to the environment. You might as well say that a thermostat is intelligent because it "knows" when to turn the HVAC system on and off. In response to gravity, the roots are just "saying" to themselves, "Grow in the direction starch grains go." They are using information from the environment and comparing it to genetic instructions, but to call this intelligence is stretching the idea too far. (However, some plant responses are complex enough that they might be considered proto-intelligence, as explained by Daniel Chamovitz.)[2]

Yes, I am saying that plants are dumb. I say this even though I am a botanist and I study plants and love them. I did not say they are stupid. To be stupid means to do something that you ought to know is incorrect. Plants are, however, dumb because they do things without knowing what they are doing. When we see leaves growing toward the sun, we tend to impute intelligence to the leaves, as if they have decided to grow toward the sun, because this is what we would do if we were leaves.

Just to say that plants are dumb is not to minimize the amazing things they can do. One day, as we walked along the river, my wife and I looked at a tree we had seen many times, but this time we looked at it more closely. We saw this time that it was actually two trees of the same species that had grown together. My wife pointed out that they were two trees, rather than just one that had split apart early in its growth, because each of the fused trunks had its own set of roots, the roots of one tree growing over or under those of the other. This pair of trees looked like one tree. But the reason was very simple: branches grow toward the light. Therefore, when two trees are adjacent to one another, the branches that are on the side of each tree *away from the other tree* are the ones that grow.

Another thing that trees do that makes some people think they are intelligent is that they can communicate. If an animal begins to eat the leaves of one tree, that tree may send a slow electrical impulse down through its roots. This electrical message can travel through fungal strands into the roots of other trees, causing these other trees to begin manufacturing poisons in their leaves. One tree "tells" another that herbivorous animals have arrived. A French news report said, "Intelligent, the trees? After seeing this report, you will not doubt it anymore."[3] This is indeed a sophisticated communication mechanism, but the trees do not actually know what they are doing.

Sometimes intelligence is not worth the price. It requires a brain, and brains are expensive. Alternatives to intelligence, such as the simple growth responses of plants, can even lead to better results. If the two trees growing together that I mentioned above had been intelligent humans, they would have had a harder time of trying to figure out where to grow their branches and roots—like two people trying to share an apartment and getting into fights. But by following a simple rule, and not thinking about it, the two trees reached a perfect solution.

INTELLIGENT ANTS?

And then there are the ants. They have an astonishingly effective alternative to human-style intelligence. Ants are one of the most mysterious topics of study in the world. They live almost everywhere, and everybody knows about them—or thinks they do. Each ant is small and not very smart, but collectively an ant colony displays what most people see as a creepy form of intelligence.

One day, before I went to work, I closed up a bag of potato chips and left it on top of the refrigerator. This was back when I still ate lots of potato chips. I thought the potato chips were pretty inaccessible to the ants. But when I came home, there was a thick phalanx of ants coming from the door, up the wall, and into the bag, through the tiny, tiny point at which the bag touched the wall. How did they do this? How did they figure out where the chips were? By scent, you probably think. And that must have something to do with it. But how did they find that little tiny point of contact between bag and wall?

Actually, it was not intelligence, as I defined it. Each ant has very little intelligence. The forager ants (as opposed to the ants that defend the colony or take care of the young) follow these three extremely simple rules:

- Wander around at random looking for food.
- If you find food, take a bite and then head straight home, laying down a scent trail as you go.
- If you find some other ant's scent trail, stop wandering and follow it to the food, take a bite, and follow the trail back home, laying down a scent trail as you go.

If a single ant finds the food, it doesn't take very long for other ants to follow its scent trail, making it into a scent superhighway that consumes the attention of all the ants in the colony. Chaotic wandering undergoes a transition into orderly foraging.[4] We can think of this as a collective form of intelligence. You could even think of the colony of ants as an organism in its own right—a *superorganism*, as Edward O. Wilson calls it.[5]

Ants therefore respond to the world in a very different way from us. A primitive hunter would think ahead about where the quarry is or will be. An ant just follows robot-like rules. It is a different kind of mental adaptation, but it works. What insects lack in intelligence they make up for by being abundant. H. G. Wells understood this when he wrote *Empire of the Ants*. Daphne Du Maurier used this idea also in *The Birds*. Birds are among the most intelligent animals, but when they formed big flocks that attacked humankind, in Du Maurier's story, they were not *planning* to destroy us. She just wrote that they were following the dictates of millions of years of evolution.

We humans tend to think of animals as being more intelligent than they really are. I am thinking of Buck the dog in Jack London's *Call of the Wild*. It is doubtful that Buck could have known all of those things London attributed to him, though London tried very hard to give Buck a different form of intelligence from humans. On the other hand, we don't want to go too far the other way. Mammals and birds have all gradations of intelligence. And they are not robots (even though individual ants may be). The difference between a mammal or a bird and a robot-like insect is that the mammal or bird can get bored doing the same thing over and over. Robots, and insects, do not.

SELF-AWARENESS AND EMPATHY

Amazing as it may seem, many scientists, including René Descartes, used to think that mammals did not experience pain. They may act like they do, but they are just going through reflexes, he said. If he lived today Descartes might have described them as robots. Only in the last century or so has science shaken off this assumption and started to treat animals as sentient beings that can, in fact, experience pain, even if they are not highly intelligent.

But we might mean something more than this for intelligence. We might want to include the capacity of *self-awareness*, something that humans and a few other species have. Many people, some of whom I am very close to, use cats as an example of a creature that has self-awareness. Anyone who owns or is owned by a cat will try to do this. I am not saying that your cat Bartholo-Meow does *not* have self-awareness. I am just saying that you cannot prove it. Bart does not necessarily love you (sorry) but just wants to get warmth and food from you, and knows subconsciously that purring will do the trick.

Nobody knows how to determine whether animals have self-awareness. But there are two ways that have been tried, both very interesting. The first is the *mirror test*. Does the animal recognize itself in a mirror, or does it assume the image is another animal? Most birds flunk this test. They see, and attack, their mirror images. In several American cities, birds have figured out that if they sit on car mirrors and bend over, they will find another bird, which they attack. This may break the mirror. The birds do not realize their error, so one bird will imitate another. Soon birds all over town are doing this, and people have to wrap mirrors in plastic bags while their cars are unattended.[6] A male cardinal attacked its reflection in my bedroom window every morning at dawn in the spring mating season for two years.

A variant of the mirror test is the *rouge test*. This is often what is meant by the term "mirror test." You put a little bit of cosmetic on a mammal's shoulder, where it cannot see it, then let it look in a mirror. If it is self-aware, it will, while looking in the mirror, try to remove the rouge. It knows the image in the mirror is itself and that it has something on its skin or hair. Now, you may ask, "How do you know they do not just feel the cosmetic and try to scrape it off?" Researchers place some invisible cosmetic base on the other shoulder, which feels the same as the rouge but cannot be seen in the mirror. The self-conscious mammals scrape only on the rouge. As it turns out, apes (chimps, bonobos, orangutans, gorillas, and of course humans) have self-awareness as measured in this way. So do orcas and bottle-nosed dolphins.[7] And, despite what I said earlier about birds, European magpies can pass the test. As work by Bernd Heinrich and others shows, crows, magpies, and their relatives are rather intelligent.[8] Apparently, crows can even remember the faces of people who have threatened them in the past.[9]

And elephants. Elephants are disturbingly intelligent.[10] I say disturbingly because their mental world is closer to ours than many of us like to think. Not only do they recognize individuals within their herds (something that many animals can do), but they also remember them. One researcher played back the sound of a recently deceased matron of the herd, and some of the younger elephants acted as though they were grieving. Does this mean that they are aware of death? Or at least do they miss the old elephant and think that she is somewhere else?

EMPATHETIC MICE?

I do not know how a scientist could answer those questions about the elephants. But one group of scientists set out to show that mice can feel empathy.[11] Let me tell you about it.

It is easy enough to show that a mouse will display signs of discomfort—that is, it will wriggle and writhe in response to seeing another mouse in pain. This was the first thing that the scientists demonstrated. They placed two mice in separate plexiglass containers that were just large enough for them to move but not for them to turn around. Then the scientists irritated one of the mice by poking it with a syringe that contained diluted acetic acid (vinegar) solution. Of course, the scientists had to invent a way of measuring how much the mice wriggled. Other researchers, working with other species of animals, use a standardized way of quantifying how much the animals frown when uncomfortable. They call it the grimace scale.[12] The mouse researchers had a similar approach.

In response to seeing one mouse wriggle, the other mouse often wriggled or writhed. (Even which term you use reveals an underlying bias about whether you think the mice are experiencing pain.) The scientists knew that it was sight that induced the writhing response because when they used opaque plexiglass tubes, in which the mice could not see one another, the response did not occur, at least in excess of the usual background wriggling that mice do in plexiglass tubes. The scientists knew that vision was the modality by which the mice communicated because deaf mice or mice unable to smell could still respond to one another's pain.

But just because one mouse writhed when it saw another mouse in pain does not mean that it felt empathy. It might have simply felt fear, from the anticipation that the same thing will happen to it, which does not require a big brain. How can you distinguish empathy from fear?

If mice form empathetic connections, they are most likely to develop them with other mice raised in the same cage—the same plastic box with sawdust bedding and water bottles, which is the world of the lab mouse. You cannot create empathy in mice that have no capacity for it, nor can you eliminate it in mice that do. But you can reasonably expect that mice raised in the same cage (cage-mates) will have *more* empathy for one another than do mice raised separately.

When the scientists performed the wriggle test, they found that a mouse was significantly more likely to wriggle in response to seeing its cage-mate poked and irritated than if it observed a "stranger" being so treated. It wasn't that the cage-mate always wriggled, or the stranger never wriggled; there was, however, a significant difference in how much they wriggled. The scientists' conclusion was that mice can develop empathy—perhaps not for all other mice but for those who grew up in the same cage.

Altruism (explored more fully in a later chapter) is something you do; empathy is something you feel, which reinforces altruistic behavior. If the capacity to feel empathy is an indicator of personhood, then mice are at least in the antechamber. I suspect all social mammals can feel empathy. Dogs certainly do; they even feel empathy for humans, as humans do for them. But cats? Cats are, relative to dogs and humans, loners. But from what I have seen, they have at least as much empathy as those mice do. Cats can get very upset when a littermate experiences trauma. I suppose this could be experimentally tested somehow. But did you ever try putting a cat into a plexiglass tube?

TELL ME WHY

Scientists resist the bias of agency. They are not satisfied with answering every question with "God did it." This is why many scientists would feel uncomfortable with the old song "Tell Me Why."

> Tell me why the stars do shine
> Tell me why the ivy twines
> Tell me why the sky's so blue
> And I will tell you just why I love you.

A scientist would be tempted to say, "The stars shine because of thermonuclear reactions described by the formula $E = mc^2$! The ivy twines because when one side of the stem touches an object, that side of the stem expands less than the other side, a process called (I am not making this up) thigmotropism! And the sky is blue because of Rayleigh scattering of photons by the atmospheric molecules!"[13] The song, however, simply says that God does all of these things. Because God made you, that's why I love you.

But, at a deeper level, why does the world exist at all? Why are there stars, ivy, and skies? Many religious scientists take refuge in the question "Why does anything exist rather than nothing?" To them, God is the ultimate ground of being. And regarding this, science has nothing at all to say.

DO IT YOURSELF!

Find an example of animal intelligence—for example, a mockingbird looking for insects just after you have mowed your lawn—and explain it as a simple reaction rather than as a reasoned response on the animal's part.

CHAPTER 11

MEASURING WHAT YOU THINK YOU'RE MEASURING

One of the recurring problems of scientific research is, How do we know that what we are measuring is what we think we are measuring? As I mentioned previously, this refers to the *validity* of scientific studies. The validity of a piece of scientific research is that it says what it means and means what it says. I have mentioned this concept briefly, and now it is time to look at it more closely.

CONSTRUCT VALIDITY

One of the most important kinds of validity is *construct validity*. Does the variable that you measure honestly represent the concept that you want to understand? Have you *constructed* a measurement that really tells you what you want to know?

This could be something as simple as air temperature. To measure air temperature, one might think, you just put a thermometer out into the air. Warmer air will cause the fluid in the thermometer to expand, which causes the fluid level in the tiny capillary tube to rise. What could be simpler? But if the sun is shining on the thermometer, the thermometer will absorb the light and become warmer than the air, in which case the thermometer is no longer measuring the air temperature. You could, alternatively, put the thermometer inside of a box. But if the box is a dark color, it will absorb sunlight and make the air inside the box warmer, in which case you are no longer measuring the temperature of the air outside of the box. To measure air temperature in a valid fashion, the thermometer should be inside of a white box that is open to the air—which is why meteorologists put their thermometers inside of ventilated white boxes. I'm sure you've seen these boxes at weather stations. There is usually a little whirling set of cups (an anemometer) on top of the white box to measure wind speed.

The measurement from a thermometer is objective, meaning that it is not influenced by the observer. But subjective measurements, based on the observer's feelings, are sometimes useful. As simple a question as "Is it cold outside?" is something that, in order to be useful to you (as in helping you decide whether to put on a coat), depends on a large number of objective measures: air temperature, wind speed, body heat production, to name a few. If your body's core temperature is a little low, your whole body will feel cold at an air temperature that you might otherwise find warm. But if you step outside to determine whether it "is cold," your brain is actually performing a complex, subconscious calculation that takes all of these factors into account. Such subjective measurements might have more construct validity than any one of its contributing factors in helping you decide whether or not to wear a coat.

Whether the outside world feels cold or not can even depend on whether you have been drinking. Alcohol causes the little muscles around the arteries in your skin to relax, which allows more blood into the skin. This makes your skin warmer, and the heat sensors in your skin tell your brain that you *are* warmer. But this comes at the expense of losing heat from your body core. Drinking when you are trapped out in the cold is actually a really bad idea. Therefore, subjective measures might have more, or less, construct validity than objective ones.

Construct validity could be challenged by something as conceptually simple as how to measure toxicity. When we say a kind of chemical is toxic, what do we mean? Toxic to what? We have to measure the effects of the chemical on the type of organism we are interested in. If we are interested in toxicity to humans, we usually measure the effects of the chemical on rats or mice, which are physiologically similar to us. But it is time-consuming and expensive to do experiments on large numbers of mice. A scientist may choose to do a preliminary toxicity study upon a simple kind of animal that can be raised in large numbers (tiny aquatic invertebrates are sometimes used) before performing the study on a smaller number of mice.[1]

The most important pieces of scientific research are those that address variables that are hard to measure. Human health is a perfect example of this. What is health? We all have a general impression about what it is, but there is no such thing as a health-o-meter, nor are there any metric units of health such as the picovigor or the megavitality. We have to substitute some measurable quantity for this general concept.

CONSTRUCT VALIDITY AND HEALTH

In studies of plants and of insects, you can validly assume that if they weigh more, they are healthier. This is because plants and insects are never, as far as I am aware, overweight.

First, consider plants. In my studies with plants, I routinely assume the biggest plant in the species is the healthiest. (Of course it is not fair to compare little grasses against big trees.) But a comparison of plant weights, even within a species, is not always valid. There was a botanist named Eckard Gauhl who researched what he considered to be sun-adapted and shade-adapted plants within the same species.[2] That is, natural selection had produced two groups of plants in this species: those that grew best in the sun, and those that grew best in the shade. He showed that the sun-plants grew better in the sun than in the shade, and that the shade-plants looked sick when grown in the sun. Now, on the face of it, this makes sense. You know perfectly well that if you expose your African violets to bright sun, they will turn brown and shrivel up. But Gauhl's "shade-plants," as it turned out, were infected by viruses. His "sun-plants" were not. His "shade plants" grew better in the shade than in the sun not because they had shade-adapted physiology but because the hot sunny conditions revealed the sickness caused by the viruses. Weight was a valid measure of health, but not the kind of health Gauhl had assumed. He was dead before other botanists discovered his error.[3]

I did a study in which I measured toxicity of different kinds of leaves by weighing hornworms.[4] Hornworms are caterpillars (of two closely related species) that eat the leaves of tomato, tobacco, and even poisonous jimsonweed plants. They are the bane of gardeners. You have probably seen them. They are as big as a finger, colored green, and have black-and-white stripes and pointy red tails. They look like birthday candles with a red wick at the end. As it turns out, you can raise hornworms in little plexiglass vials and feed them a special hornworm chow. That is, they do not have to eat tomato or tobacco leaves. Inside of a vial, a small hornworm grows into a large one, eating up the chow. All you have to do is to scoop out the waste products once in a while. And you can weigh them as they grow. To do this, you wash them, dry them, and put them on the balance, recording the weight before they crawl off. This means you have to handle them gently, even while they are chewing on your fingers (they cannot actually bite through your skin). Am I creeping you out? If not, maybe you could be a scientist.

So in one of my experiments, I wanted to compare the toxicity of material from two different plant sources. These were not plants the hornworms would ordinarily eat, but if you grind it up and mix it in with their chow, they eat it anyway. I found that the more toxic plant material was the one that made the hornworms grow more slowly and reach a smaller final weight. It also turned them from a bright green into a sickly blue color. Heavier hornworms were healthier.

But you cannot assume that a heavier mammal is a healthier one. A healthy person has a weight that is somewhere in an optimal range, not too heavy, not too light. In order for weight to have construct validity for humans, it needs to fall within that range. We use optimal ranges for many different kinds of health measurement: blood sugar levels, blood pressure, white blood cell count, visual acuity, etc. There is no single optimal measurement; it is a range that includes normal population variability.

CONSTRUCT VALIDITY AND THE ECONOMY

While it is not valid to use weight as a measure of health in a mouse or in a human, we routinely measure the health of the economy by measuring its grossness, by using the aptly named gross domestic product (GDP). But a large GDP (like the closely related gross national product or GNP) does not necessarily indicate a healthy economy, just a large one. A large economy, like that of the United States, which consumes a lot of resources—more than the earth can continue to produce—and produces a lot of waste products is not necessarily a healthy one. If you have a lot of sick people smoking cigarettes and paying big hospital bills, or having someone else pay their bills for them, this might contribute more to the GDP than do a lot of healthy people who don't go to the hospital and don't buy cigarettes. Staying home and cooking your own healthy meals contributes less to the GDP than does going out to eat big unhealthy meals. Drinking iced tea at home contributes less to the GDP than consuming syrupy drinks at a fast-food place. As economist Paul Hawken said, "We have an economy where we steal the future, sell it in the present, and call it GDP."[5] I maintain that the GDP is not a valid measure of economic health. As another author said, you can't eat GNP.[6]

Another measure of economic health is how evenly the wealth is distributed among a country's citizens, rather than the total amount of wealth. In Japan, the average CEO of a large corporation earns sixty-seven times as much income as the average employee of the corporation. In France, CEOs earn 104 times as much, and in Germany, 147 times as much. In the United States, CEOs of large corporations earn 354 times as much.[7] Those who reject the construct validity of the GDP insist that this is a problem. We have a lot of wealth, but the extremely unequal distribution is an indicator that our economy is not healthy. The Gini coefficient is a measure of evenness with which data values are distributed about the mean, with a value of zero for a completely uneven distribution, and a value of one for a sample, or a society, in which all people are totally equal.[8] Evenness of wealth distribution is not the only thing that most economists want to know about a country, but neither should they be solely interested in the grossness of the economy.

The Dow Jones Industrial Average (DIJA) is a more valid economic indicator, not of the actual health of the economy but of how people feel about it. It is a scaled average of stock prices. A one-dollar increase in the average stock price results in an approximately seven-point increase in the DJIA. A high "Dow" means that people are investing more money in stocks, which means they have money and they have confidence. The Dow can fall if people *feel* as if they cannot trust the companies in which they would otherwise invest. True, the Dow is a subjective measure, but it is that subjectivity that we may want to know about. It does not prove that an economy is actually healthy but just that people think it is. Of course, it may reflect what former Fed chair Alan Greenspan called "irrational exuberance."[9]

The unemployment rate is often used by politicians as an indicator of economic health. You hear it nearly every day on the news. But, it is subject to extensive manipulation. When unemployment decreases, politicians take credit and claim they are job creators (figure 11-1). But when unemployment increases, these same politicians will say that this is the result of more people looking for jobs, since unemployment statistics do not count those who have "dropped out of the labor force." How useful is a measurement that can be interpreted any way you like it?

Figure 11-1. How good is the construct validity of the employment rate as a measure of the health of an economy? (Cartoon by Leslie Gregersen.)

There are certainly measures of economic health that are less valid than the GDP. The government of North Korea appears to use the number of missiles in its arsenal as a measure of economic health, even as the people are on the brink of starvation.[10]

Maybe we need a new kind of measurement for societal health that is not directly linked to economic grossness. In 1972, the new king of Bhutan, Jigme Singye Wangchuck, opened up what had previously been a kingdom almost entirely closed off from commerce with the rest of the world (it was probably the inspiration for Shangri-La). But he rejected the idea that the GDP was a valid measure of his country's success. Instead he proposed the GNH, or gross national happiness. Bhutan might have one of the lowest GDP levels in the world, but, the king said, it sure is a nice place to live. (A cynic would say that he might also have wanted to use the GNH to convince the citizens that they ought to be happy and not cause any social disruption.) At first the GNH was a very unclear (and typically Buddhist) concept, as was admitted even by Buddhist scientists. But over the years it has been refined. For example, a 2006 version of the GNH was proposed by the International Institute of Management.[11] It used surveys and societal data to measure the following things:

- *Economic wellness:* things such as low consumer debt and everybody having an income that is high enough to buy what they need
- *Environmental wellness:* things such as freedom from pollution, noise, and traffic
- *Physical wellness:* things such as low levels of severe illnesses
- *Mental wellness*: things such as low levels of antidepressant use
- *Workplace wellness:* things such as low unemployment and few lawsuits
- *Social wellness:* things such as low divorce and crime rates
- *Political wellness:* things such as the functionality of local democracy

Note that nearly all of these things can be objectively measured, as well as subjectively measured by surveys that indicate how people feel about them. I know you can quibble about each of them. Low divorce rates might mean that women are oppressed, and low levels of antidepressant use might indicate a shortage of medicine. But the GNH is well on its way to being a more valid measure of how good a society is to live in than the GDP ever was or can ever hope to be.

Many people in France think that we in the United States are crazy, mainly because of our abundance of firearms. Some imagine that America must be like the Old West, where everybody walks around with a six-shooter. But what measurement, with construct validity, can they use to justify this impression? There is no single measurement, but a reporter for a French news agency used three numbers to make his point: the high number of gun-related deaths in America; the high number of mass shootings (about one per day); and the fact that, while 78 percent of Americans do not own guns, 3 percent of Americans own half of all the guns. His source defined mass shootings as four or more deaths or injuries from a single gunman at a single incident.[12] These three figures, taken together, have more construct validity than would any one of them would have separately.

CONSTRUCT PSEUDO-VALIDITY

A commonly used measure, the construct validity of which may be questionable, is the Google search rank. At first, I was pleased when I saw that my blog showed up on only the second page of a search I did on a certain general topic. Then I realized that Google probably weights the ranks in favor of websites recently visited. Of course, I have to visit my own blog frequently, to post on it. My pleasure was short-lived.

Another example of questionable construct validity is the barking of dogs. Many people have dogs to protect their homes. A dog may bark at anything, not just an intruder, and may keep barking long after the initial stimulus has passed. Not very useful information to use in alerting one to the presence of a criminal. But what if a dog does *not* bark? If a crime takes place and the dog does not bark, it might mean that the criminal is someone well known to the dog. The failure of the dog to bark has some construct validity as a way of identifying the criminal. This is the basis of a famous Sherlock Holmes story, "The Adventure of Silver Blaze."

Generally speaking, the more complex an index is—the more kinds of measurements that go into it—the riskier is its construct validity. Magazines and websites frequently publish lists of which cities are best, or worst, to live in. Seldom do the lists even come close to agreeing. For example, a recent list of the ten worst cities in America by one website put Tulsa, Oklahoma, at the top of the list. I live in Tulsa, and it has its problems, but this ranking surprised me. Meanwhile, another website listed Detroit as the worst city. The ranking depends entirely on the sources of the data and the relative weights assigned to them. In view of the subjectivity of such rankings, they are practically worthless as general indicators. Perhaps the best thing for a person to do, when considering where to live, is to decide for himself or herself what is most important and then search for the data regarding those things. If you want to live in a city with the fewest murders, stay away from Baltimore.[13] If you don't like cities with high poverty rates, stay away from Brownsville, Texas.[14] If you don't like air pollution, you should stay away from Visalia, California,[15] rather than New York City. These, not the composite measures, are the ones with construct validity that you care about.

Probably the source of data with the worst imaginable construct validity is testimony obtained under torture. Torture has been used throughout history to get information from captives. The United States claims to not use torture, but this depends very much on the definition of the word. The George W. Bush administration defended the use of "enhanced interrogation techniques."[16] Humanitarian considerations aside, what construct validity does it have? I know that if I were tortured I would admit to anything—even to being a space alien. In 2014, the CIA admitted that its previous use of torture had provided relatively little information that was useful for national security.[17] In 2017, President Donald Trump reversed the recent trend of the American government away from torture, and proclaimed that "torture works."[18]

EXTERNAL VALIDITY

I will briefly mention one other kind of validity. A measurement has *external validity* if the measures made upon a set of samples can be generalized to the outside, or external, population. Remember that I previously wrote that a sample must be a valid representation of a population. The conclusion is only as valid as the extent of the sample. This is the most obvious example of external validity.

Take, for example, a study I conducted with my colleague Erica Corbett.[19] We wanted to know how much insect damage occurred on the leaves of post oak trees in south central Oklahoma and, if possible, to explain any patterns that we saw. We needed a sample with as much external validity as we could reasonably get. We took leaves from twelve trees because one or two trees could certainly not represent all of them. We took the leaves each month during the summer, from May to September, since insect damage might be higher at one time than another. We took leaves from these trees for five years because each year might be different. In fact, some of our samples came from a drought year, a year with moderate rainfall, and a flood year. We had 2,574 leaves. That was a lot of work, but anything less than that would have not been generalizable to more than one tree or more than one year.

A young scientist recently challenged his older colleagues to reexamine the external validity of studies of animal nervous systems. He pointed out that the vast majority of recent neuroscience papers focused on rats, mice, and humans.[20] He said he just wanted to understand why such a limited range of species was studied. Perhaps stretching a bit, he compared himself to the boy in the Hans Christian Anderson story "The Emperor's New Clothes." In the past, the scientist pointed out, neuroscientists got important insights from studying squid (which have enormously thick nerve fibers), frogs, and horseshoe crabs. Perhaps today we are missing out on some important discoveries by limiting our external validity. Moreover, many psychological studies are performed upon university undergraduates. This is because many psychological researchers teach at universities, and they can recruit students for tests by paying them little, or perhaps paying them only with extra credit. But, since university undergrads are not completely representative of society, these studies also have limited external validity.

One of the biggest examples of the problem of assuring external validity is the science of global warming, or global climate change.

Climate is not weather. Climate is the long-term average of weather.

So you can't just stick your head out the window and know whether global warming is occurring or not. Politicians who reject global warming, such as Oklahoma senator Jim Inhofe,[21] stick their heads out of the window in winter, notice that it is cold, and wonder, *Where is global warming now?* (There are two answers to this question. First, the warming is somewhere else. During the cold winter of early 2014, as experienced in North America, the heat was so excessive in the Australian summer that the shoes of tennis players were melting on the court.[22] Second, the warming may be gone at the moment, but it will be back.) Similarly, an environmentalist may stick his or her head out the window in summer and say, "It's so hot! It must be global warming!" To say a hot summer day proves global warming is just as invalid as to say that a cold winter day disproves it.

To test the hypothesis of global warming, you need to know global average temperatures. This is not an easy thing to estimate. You need to have a globally representative set of temperature measurements. While we have thermometer readings from Europe and North America before the nineteenth century (for example, Thomas Jefferson had a newfangled thermometer at Independence Hall as the Continental Congress met), we have no measurements from Africa, Asia, South America, the Arctic, or the Antarctic from that time. But starting in the middle of the nineteenth century, British and American navy ships all over the world began recording daily temperatures. At last, this kind of global coverage made the estimates of global temperatures more externally valid. This is why the time axis of the typical graph of global temperature begins after 1850.[23] Earlier temperatures can be estimated, using such "proxy" measures as tree rings and ice layers, but these are less accurate than actual thermometer readings.

But even temperature measurements by worldwide navies do not completely satisfy the necessity for external validity. Yes, ever since the beginning of Arctic and Antarctic research, scientists have been measuring temperatures in those regions. But, for understandable reasons, there are fewer temperature readings from the North and South Poles than from places where more people live. This introduces a bias in favor of temperate and tropical regions. This bias would tend to underestimate, rather than overestimate, the amount of global warming that has occurred. Fewer still have been the measurements of deep ocean temperatures, producing a bias in favor of land and ocean surfaces. Scientists were puzzled that, although carbon dioxide concentration continued to increase in the atmosphere, global warming seemed to be slowing down after about 1980, before speeding back up. As it turns out, the Arctic regions have been warming rapidly,[24] as have the deep ocean waters, since 1980. So the earth has continued warming, but more in the Arctic regions and deep oceans than in temperate and tropical regions. The underrepresentation of Arctic and deep ocean temperatures was an external validity problem that caused temporary confusion in the science of global warming.

CONCLUSION

One could say that the process of science is just a formalized way of trying to think of all of the things that could go wrong in research—such as bias and invalidity—and to try to correct for them. Scientists spend a lot of time sitting around, or hiking, or drinking, while they try to think of how their measurements, in which they are prepared to spend a lot of time, money, and emotion, might not be valid. And, as shown in the case of poor Eckard Gauhl, the truth might come out too late.

DO IT YOURSELF!

Think of something important to you—your personal economy, or your health—and evaluate the validity of the information that you use to determine it.

CHAPTER 12

OOPS, I HADN'T THOUGHT OF THAT

Sometimes, despite our best efforts, things happen that we had not thought of, and which may totally undermine the reliability of our scientific conclusions. I will just give a few examples.

The attrition effect. As an experiment goes along, whether with plants or people, there is some attrition. Some of the plants die. Some of the people drop out of the study. The death of the smaller plants or the departure of the less motivated or less capable people will boost the average quality of the remainder. Consider a professor who teaches a course. He may give an exam at the beginning of the course (the pretest) and then give the same exam at the end (the posttest). This is standard procedure, actually. If the average posttest score is higher than the average pretest score, he draws the eminently reasonable conclusion that the students have learned something from his course. But it could just as easily mean that the students who could not handle his course dropped it. The students at the end may not have actually learned anything; they may simply have persevered. In order to evaluate whether the students have learned anything, the dropouts should be eliminated from the average score on the pretest as well as the posttest. Have you ever mistaken perseverance for improvement? I've been there.

The sequence effect. Suppose you wish to assess the mathematical and verbal abilities of students and you give them a math test and a reading test. They will do better on whichever test they take first. They get bored or tired during the second test. Unless the first test is at 8:00 a.m. Then they might bomb both of them. The solution is as simple as it is obvious:

give half of the students the math test first and half of them the reading test first. While this will not necessarily help you to *evaluate* any individual student, it will help you to *assess* the student body as a whole because you can average out the sequence effect. I've been there too.

The history effect. The way each plant, animal, or person behaves in the experiment depends on his or her or its history, where it has been before and what has happened to it. Because they have *so* been there. Within any group of people with whom you are working, you have older people (who may be more experienced with life) and younger people, and people with different levels of prior education. I always get classes of students, some of whom do not know how to use a ruler (I am not making this up) and some of whom are astonished at how primitive our lab equipment is. The history effect is why scientists cannot reuse mice from one experiment to the next. There can be a carryover effect from the drug used in the first experiment into the second. So for each experiment, a pharmaceutical researcher needs to get rid of the old mice and use a fresh batch. We'll just leave it at that. I've been there.

The motivation effect. This mostly applies to humans and other animals. For a rat to figure out a maze, it has to *want* to figure out the maze. We run into this problem with student assessment all the time. We have to evaluate their skills individually (to calculate their grades) and assess their skills collectively (to demonstrate to the administration that our students are learning something and therefore getting something for their money). You might ask what the problem is: they want good grades, so of course they will try to do well on tests. But the assessment tests that we give them do not perfectly match our courses. This is because the assessment tests are standardized for the whole country and for different types of courses. This increases their external validity. Despite the fact that the assessment test is not a perfect evaluation instrument, we go ahead and count the posttest as part of their grade; otherwise they would just come to the test, answer "ABCDEABCDEABCDE" on the answer sheet, and leave. (I've seen this happen.) Of course, even the promise of good grades (or the threat of bad ones) may not be enough to motivate students, who are still young enough that only immediate rewards and punishments matter to them. The prospect of a bad grade transcript (which will follow them all of their lives) may seem distant to them.

Surprise factors. Just when we think we have thought of everything there is to think of, some surprise factor comes along that at best introduces random error, and at worst introduces bias, into our studies. In 2014, scientists demonstrated that lab rodents experienced stress when handled by male researchers.[1] They eventually figured out that this effect was caused by some odor that the male researchers produced, and occurred even if the odor was simply left on the sawdust bedding of the rat cages. It is safe to say that, prior to this research, nobody had suspected that this might occur. The only way to avoid such errors is to try to have all factors distributed among all treatments. For example, you would not want a female researcher to handle all of the rats in one treatment and a male researcher to handle all of the rats in the other. If both researchers handled rats from both treatments, the gender-odor effect would be evened out and possibly not have any net effect on the results.

Often, by a simple modification of experimental design, we can avoid these attrition, sequence, history, motivation, and surprise factor effects. And sometimes not. But at least we should not let them take us by surprise.

Then there is the Sherlock Holmes effect I mentioned earlier, when people assume that their minds have encompassed all possible alternatives. This is the error on which almost all of creationism—and its suave cousin intelligent design—is based. Their basic argument can be summarized in this way: "Look at this very complex system! I cannot imagine how it could possibly have evolved, therefore it did not." Evolutionary scientist Richard Dawkins calls this the "argument from personal incredulity."[2]

One of the founding documents of intelligent design, Michael Behe's *Darwin's Black Box*, makes numerous errors of this sort.[3] Behe's main argument was that the natural world shows numerous and pervasive examples of *irreducible complexity*—that is, complexity in which the removal of even a single component would cause the system to collapse. He chose one of his examples to illustrate his point about complexity: a mousetrap consisting of five components. Remove any component and the trap will not work. What he was unable or unwilling to consider, however, is that simpler mousetraps can indeed exist. One person even designed a mousetrap consisting of a single piece.[4] And of course there are other and simpler ways to accomplish the same thing as a mechanical mousetrap with a hammer—sticky traps, for example. While it is true that a mousetrap designed to use five components cannot work if you remove one, simpler designs are possible.

Another of Behe's examples is the blood-clotting system found in complex animals. It involves a series of proteins, in which one protein acts after another. Remove any one protein and the system fails. This is what happens in the blood disease hemophilia. Behe could not imagine how all those proteins could evolve, nor could he imagine a simpler system. But, in fact, different animals do have different systems, and blood clot formation is simpler in many animals than in humans. And many of the proteins are just modified copies of the serine protease enzyme; the clotting proteins did not each have to be designed from scratch.

This anti-evolution argument is not new. The vertebrate eye, for example, is so complex that Darwin himself admitted that he could not imagine how it might have evolved.[5] For over 150 years creationists have gleefully referred to this problem. Of what use, they claim, is half of an eye? The answer is simple. An eye that is half as efficient may be half as useful as a full eye. An eye only 1 percent as efficient as ours may be infinitely more useful than having no eyes at all. Having imperfect vision is better than being blind. As the old adage goes, "In the country of the blind, the one-eyed man is king." Richard Dawkins made a humorous observation about this point.[6] He referred to a certain colleague who so seldom cleaned his glasses that his vision could have been little better than a blur anyway.

How can one, in fact, demonstrate that eyes simpler than ours can, in fact, function? The most direct way is to find such eyes in the natural world, "designed" by natural selection. Some aquatic animals have simple "eyespots" that are sensitive only to the presence or absence of light. They can do little more than inform the animal that a shadow has passed overhead. But if that shadow happens to be a predator, it can be useful information. Some single-cell organisms have eyespots that inform them which way to orient themselves, since light comes from overhead. Some animals have eyes that are like camera apertures, little more than a muscle-controlled hole and no lens, which allow blurry images. The best eyes have lenses, which focus the light, but there is a whole range of complexities found in lens eyes. Darwin gave some of these examples in *The Origin of Species*—examples that creationists have studiously ignored for a century and a half.

We are all vulnerable to this sort of thinking. In my own case, I find myself unable to believe in the theory that there is an infinite number of universes. My skepticism is entirely emotional—it simply boggles my mind. But, were I to ever discuss my feelings with an infinite-universes proponent such as British Astronomer Royal Sir Martin Rees, he would rightly ask me, "What part of 'infinite' do you not understand?"

DO IT YOURSELF!

Look for examples of the sequence effect, the motivation effect, etc., in your work.

CHAPTER 13

EVERYBODY'S BIASED BUT ME

One of the greatest sources of prestige for scientists is that our research is considered to be unbiased. This is why our statements carry more weight than those of politicians—not just because we are less likely to be dishonest (see the next chapter) but because we try our best to eliminate bias. Politicians recognize this. Even though each politician trumpets his own unimpeachable honesty, he also scoffs at the others. Therefore, politicians sometimes establish "nonpartisan" bodies to provide them with an unbiased summary of research into important questions.

An important example is the Congressional Research Service (CRS),[1] a branch of the Library of Congress. When, for instance, Congress wanted to know in 2012 whether increasing taxes on the wealthiest Americans would harm the economy, they turned to the CRS. The resulting CRS report concluded that slightly higher taxes on the wealthy would not harm the economy. (They had to pay attention to the construct validity of the way they measured harm to the economy, of course.) What Congress does with this information is up to them. The Republican leadership of Congress chose to suppress this report when it first came out.[2]

Lest I make it sound as if I am dismissing bias as a moral flaw, peculiar perhaps to conservatives, let me set the record straight. Humans are always biased. All humans. Always.

One of the most important differences between a scientist and, say, a preacher is that the scientist is *aware* that he or she is biased and will *take active measures* to avoid bias. While a preacher might say, "The Lord spoke to my heart and said that we need to raise money so that I can have a private jet," a scientist would feel obligated to give you evidence of her statement that you can independently verify. As I said in the introduction to the book, science tests hypotheses, using external information.

Much of science is an organized way of recognizing and eliminating bias. As I said before, science is a discipline. It is like a yoke by which oxen pull the cart of human knowledge forward. Oxen are prone to wander, to the right or to the left, but the yoke helps to prevent them from wandering as their biases might dictate.

Measurement techniques themselves can introduce bias, even if the experimenter does not have bias. Many data sets regarding bird migrations come from weekend bird-watchers. There simply aren't enough full-time bird-watchers to get all the necessary data. But, think about it, if full-time vs. weekend bird-watchers record arrival times of migratory birds—the date on which they saw the first scissor-tailed flycatcher in southern Oklahoma, for example—the dates from the weekend bird-watchers will be later than those of the full-time watchers. If a bird arrives on a Tuesday, as many do, the daily watcher will note the Tuesday date, but the weekend watcher will record it as the date of the following Saturday.

PARTICIPANT BIAS

Participant bias affects the reliability of the information provided by the participants in the experiment. Take, for example, stress. How do you find out if participants are experiencing physiological stress? Well, you could ask the person if he or she feels stressed. But this is subjective and biased. For example, the person may believe that he should not be stressed about something, so he may report a lower level of stress than he actually feels. Scientists prefer measurements that are unbiased. A scientist might measure the amount of cortisol, which is a stress hormone, in the participant's saliva. High cortisol levels indicate stress. Cortisol doesn't lie, nor does it have a biased expression of itself.

Humans also feel better when somebody treats them for whatever malady they have or think they have. In the old days (or even today, with faith healers and exorcists) this meant having someone pray for you, lay his or her hands on you, or cast demons out of you. This is mostly bias because these practices cause the brain to release chemicals that make us feel better. For example, exercise (or perhaps an exciting ritual such as exorcism or fire walking) causes the brain to release endorphins, which make us feel better and may even make us less sensitive to pain.

Nor are these brain and hormonal processes totally without medical effect. Cortisol dampens the immune system; this is one way that stress can put you at increased risk of getting a cold or the flu. Stress can also make you careless about health maintenance. Reducing stress can, therefore, make you healthier.

For centuries, people have often felt better when they have consumed medicines, whether they were shamans' philters or modern pills and shots. This effect is so strong that in up to half of the time people can feel better when they receive a neutral material but believe it to be "real" medicine.[3] The neutral material might be a sugar pill that looks like a "real" pill. The neutral pill is called a *placebo*, and its influence on our minds is known as the *placebo effect*.

The flip side of the placebo effect has been called the *nocebo effect*. In this case, patients report annoying side effects rather than benefits of a medication, even if they are receiving a placebo.[4]

Dealing with the placebo effect has turned out to be a major challenge for medical research. It is obvious that, in testing a new drug, pharmaceutical researchers must not let the trial participant know whether he has received the actual drug or a placebo. The drug is considered effective only if it helps the person *more than* does the psychological effect of the placebo. But people are not stupid, and they can often recognize a sugar pill when they take one. They know that most drugs have a bitter taste. If the pill does not have a bitter taste, they will know that it is probably a placebo and will perhaps even feel worse as a result of knowing that their pill is not the real drug. A pill that tastes bitter may, in turn, make them feel better.[5] The use of a placebo, without the participant knowing whether she receives it or not, is known as a "blinded" experiment.

And it is not merely the appearance or taste of the pill that can invoke participant bias. It is well known to drug researchers that an expensive placebo will make people feel better than will a cheap placebo. Participants in the experiment think that if it is expensive, it must be a real drug.[6]

Sometimes bias can result from seemingly unavoidable experimental limitations. In drug trials, for example, the researchers just have to trust that the experimental subjects will take the pills that are given to them. The researchers cannot make them do so. It is possible that the only effect of the failure of some participants to take their pills is that the sample size is reduced. This would not present a bias problem. But suppose that one of the groups was more negligent than the other group. For example, the patients that received the real medication felt better and kept taking the pills while the patients receiving the placebo stopped taking pills that seemed to have no effect. This would introduce bias. One solution that has been suggested is for the pharmaceutical corporations to insert a microchip in each pill. This microchip can be traced: did it stay in the bottle, or did it proceed through the patient and out into the ecosystem? This information would at least alert researchers about patients who fail to take their experimental medications.[7]

In addition, some people respond to placebos more than other people do, just as some people are easier to hypnotize than others. Perhaps they are more vulnerable to suggestion and self-deception. On the other hand, some people do not respond to placebos at all. There appears to be a genetic basis for these differences. Some pharmaceutical corporations are thinking about identifying suggestible people and excluding them from drug trials. By doing so, they believe, they can reduce the number of participants in the trial, which will considerably reduce the cost and may increase the accuracy of the trials.[8]

One famous example of a placebo is the curative peanut oil invented by George Washington Carver. Carver is my favorite scientist, for reasons described in chapter 20. But when a lame boy with polio used some of Carver's special peanut oil and then got up and walked away, it was almost certainly a placebo effect. The boy probably had the ability to walk but felt too depressed to try. He expected to be able to walk after receiving the oil treatment, so he tried walking—and succeeded. Notice, however, that the placebo did in fact work, even if its main effect was inside of the boy's mind. This is especially true since the great George Washington Carver applied it to the boy himself.

Bias can also affect how a participant perceives flavor. Food scientists run into this problem all the time. People enjoy food more if it looks good. For example, people think that redder watermelons must taste better than watermelons with paler flesh. Trial participants may therefore receive the food while blindfolded, or in special lighting that misguides them as to the actual color of the food. The food scientists wish to know if consumers find that the product tastes better no matter how it appears to them. People also enjoy food more if it is presented well. Whole industries are based on bias. A good chef can make almost anything taste good.

In addition, people who take vitamin supplements might have a different view of themselves than the people who do not. They think of themselves as healthy. If they get a cold, they may tend to interpret it as "merely a sniffle." People who do not take vitamin C might view themselves as unhealthy and thus interpret their cold as a full-blown cold. That is, both groups of people could (objectively) experience exactly the same severity of cold but (subjectively) interpret it differently. This, in addition to the problem of correlation vs. causation, is a major reason that few scientists believe that vitamin C prevents or cures colds.

Nor do only humans experience participant bias. Scientists who work with animals in the field sometimes have to trap them, measure them, and then release them. This process is specially designed to not hurt the animals. Food lures them into a cage with a one-way door. The first time this happens to an animal, it is a harrowing experience. Maybe the second time too. But after a while the animal learns that, when it sees a trap like that, there is food inside and it won't get hurt. The animal becomes what field biologists call "trap-happy." This will not necessarily invalidate the measurements that the researcher makes upon the animal. In fact, a relaxed animal may be a more "natural" representative of its population. The problem is that it causes the researcher to sample the same animals over and over rather than a new, random sample each time. There are ways to partially compensate for this but only if the researcher is aware that it is happening.

You might ask, "What is wrong with giving drugs to patients even if it is only the placebo effect that makes them feel better?" There are several reasons. First, drugs have side effects that sugar pills do not. Second, drugs are more expensive than sugar pills. Third, the overuse of drugs can cause them to become ineffective in the population as a whole. This happens most often with antibiotics. The overuse of antibiotics can cause bacterial populations to evolve resistance to the antibiotics. If thousands of doctors give antibiotics to distraught kids (brought in by distraught parents) who have viral infections (against which antibiotics are useless), then you can end up with lots of antibiotic-resistant bacteria. This is in fact happening and may be the major medical crisis of modern times.

EXPERIMENTER BIAS

In many experiments, not only must the patient be "blind" but the experimenter must also be "blind" in order to avoid *experimenter bias*. This is called a *double-blind experiment*. If the nurse who gives the pills to the trial participant knows whether they are real or are placebos, he or she can send subtle subconscious messages. The participant picks up on the enthusiasm or lack of it and perhaps subconsciously suspects whether the pill is a placebo or not. Obviously, somebody (the lead investigator) has to know which pill is which, but once the pills are identified only by number, they are dispensed by research assistants who do not know which is which. For another example, let's go back to our food-tasting experiment. The participants may like the food better if they see the smiling face of the food scientist. Therefore, the food scientist may hand them the food sample through a little door, which prevents the participants from seeing her face. Or the participants could be blindfolded.

Before a scientific paper is published, other scientists review it and look for flaws that might indicate the paper needs to be revised, or that it is unacceptable. This is supposed to be an objective process, free from personal egos and disputes. Many journals try to hide the identities of both the reviewers and the authors, thus making the process double-blinded. Frequently, however, this is not the case. The reviewers are anonymous, but they often know who the authors of the paper are.

From behind this cloak of anonymity, reviewers sometimes vent their personal abuse on the authors. This may occur even more frequently in anonymous reviews of grant applications. While such personal attacks are far less common in science than among business people or politicians, they still occur. I am aware of no studies that determine how common such unfair attacks are; an informal poll among my colleagues indicates it occurs in about one out of five paper or grant submissions. A major motivation for such attacks is the increasing competition among scientists, who are tempted more now than ever to attack competitors.[9]

BIAS IN MARKETING

Avoiding bias is not just important in science. It is essential in everything we do. It is important, for example, in marketing. Marketers conduct research. They want to know the most effective way to get people to buy their products. The most important variable that they measure is sales. They make the reasonable assumption that if people buy it, then it is what people want. The construct validity of this assumption seems unassailable.

But it is not always true. Here is an example. I hate to buy things that are overpackaged. You know, things packaged in plastic that is so hard that you need a Klingon bat'leth sword to open it, and you may damage the product inside before you finish. Frequently, if you wish to buy a certain kind of item, you have no choice but to buy the highly packaged version. But sales of that item are not an indicator of consumer approval of the packaging. My purchase of such an item registers as an approval when in reality it is a resignation. Having no other choice, I give up and buy it. It is an example of a treatment without a control, since the consumer has no minimally packaged alternative choice.

This seems harmless enough, but overpackaging can manipulate consumers into being wasteful. As such, it contributes to the production of excess solid waste and greenhouse gases. And while it may be profitable to the corporation that sells the overpackaged item, it wastes consumers' money.

WE CANNOT THINK RANDOMLY

There is another bias that affects both participants and experimenters. I have used the word "random" a lot. But the human mind cannot think randomly. We are always noticing patterns that may not really be there, a process called *pareidolia*. The best example is the stars. The distribution of stars in the sky is random, but we see constellations. Constellations are useful for locating stars, whether you are a professional astronomer or a backyard telescope viewer. But they also seem real to us.

Researchers try to get around the bias of nonrandomness by doing mathematical analyses; computers really can "think" randomly. But when it comes to experimental design, we may think we are imposing the treatment "randomly" when in reality we are following a subconscious pattern. This is why researchers often let computers assign participants randomly to treatments or to the control.

BIAS AGAINST NEW IDEAS

Scientists have another kind of bias, but perhaps there is a good reason for this one. Scientists are biased against new ideas. But who isn't? We are more likely to reject a new hypothesis that shakes our comfortable assumptions than an old one that does not, even if the new hypothesis has better evidence to support it. As Thomas Kuhn explained, scientists cling to old notions despite accumulating evidence against them until, in a relatively short period of time, the scientific community undergoes a "paradigm shift."[10] Science, not unlike religion and politics, has revolutions of thought. But is this a bad thing? Old ideas may have been tested many times by a lot of data. If a new idea, supported by just a few data, comes along, scientists are naturally skeptical that it might just be a fluke. Even politically liberal scientists are mentally conservative. Extraordinary claims, they say, require extraordinary evidence.

One good example of an extraordinary claim is the hypothesis, previously mentioned, that an asteroid hit Earth, an idea now widely accepted. It is one thing to find a layer of rock enriched in iridium. But if the hypothesis is correct, sixty-five-million-year-old rocks *around the world* should have high levels of iridium. Researchers indeed found this iridium-rich layer in Italy, Spain, and Montana.[11]

A couple of examples of extraordinary claims come from rock carvings in Oklahoma. The first example is a rock face near the town of Heavener that has recognizable runic inscriptions.[12] The conclusion seems obvious: the rock was carved by Vikings. This is an extraordinary claim and has been much maligned. Many skeptics claim they were carved as a prank by somebody who knew just enough about runic inscriptions to get them almost, but not quite, right. Defenders of the Runestone authenticity claim, however, that the carvings were well known to the Choctaws who arrived during federally enforced resettlement in the 1830s. Rumors have also been passed down about specific individuals who admitted to perpetrating the prank. But, as it turns out, there are lots of other rock carvings in Oklahoma. Many are doubtless pranks, but that does not mean the Heavener Runestone is one of them.

The Heavener Runestone is an extraordinary claim, but just how extraordinary is it? The fact that there is no other clear evidence of Viking presence south of Newfoundland argues strongly against a Viking presence in what is now Oklahoma. How could Vikings have fought their way down to what is now Oklahoma, and, perhaps more importantly, why? Pretty extraordinary. Or is it? During late Viking times, Native American tribes had trade networks all over eastern North America. A couple of Vikings might have run away from the others, hitched a ride on one of the trade routes, and ended up in Heavener. In consideration of this, the claim may not be quite so extraordinary.

Much more extraordinary is the claim that ancient Egyptians came to Oklahoma. A carving on a cave wall (in a secret location on private land) seems to represent a horse or coyote. But some observers claim that it represents the Egyptian jackal-god Anubis. Besides the fact that the carving (according to a photograph) looks less like an Anubis hieroglyph than the Heavener carvings look like runes, there is no evidence that ancient Egyptians had seaworthy craft or ever made long voyages.[13] While they were at it, the boosters of the discovery claimed that various carved holes around the animal represented constellations. In this case, then, the claim is too extraordinary, and the evidence too ordinary, to merit belief.

And then when it's all over and the extraordinary new idea is acceptably proven, we act as if we have always believed the new hypothesis. In 1967, people scoffed at the endosymbiotic theory of cell origins proposed by Lynn Margulis, which she published only after numerous rejections.[14] This theory claims that complex cells resulted from the merger of smaller bacterial cells. Today it feels as if we scientists have always believed it. There may therefore be two stages to the acceptance of new ideas. The first stage is to say that it cannot possibly be true. The second stage is to say that it is so obvious that it need not be said.

PROVING THAT HORSES AND DOGS CAN COUNT

Our inability to recognize randomness leads directly to perhaps the most important kind of bias of all: *confirmation bias*. When we look at the world, we see some things that confirm what we already believe and some things that contradict it. Naturally, we notice more the things that confirm what we already believe. We cite evidence in support of our beliefs and ignore evidence against them, not because we are all liars but because of the instinctual bias inside of our brains. Scientists often call this bias "cherry-picking" because we pick the red cherries that confirm our predilections and fail to see the green ones. This is perhaps the main reason that scientists use statistical tests to determine the significance of the results.

I have already introduced you to one example of a confirmation bias. I used the correlation between vitamin C supplements and avoidance of colds as an example of a spurious correlation. But I also mentioned that bias may have played a role: the people who took the supplements interpreted their symptoms as sniffles rather than as colds.

In an earlier chapter, I discussed the human tendency to impute intelligence in organisms that may not really have it. A pet owner tends to think his dog is the smartest dog in the world or that her cat really loves her. This is a form of confirmation bias. Here are a couple of humorous examples.

In the early twentieth century, there was a horse named Clever

Hans.[15] His owner claimed the horse could count. People would gather around to watch the horse count. Now, horses can't talk (aside from Mr. Ed and Francis the Talking Mule), so when someone would tell Clever Hans to count to five, he would stamp the ground with his hoof five times. Clever Hans seemed to always get it right. I'm sure you have already guessed what was happening. If they told the horse to count to five, the observers would stand around calmly during the first two or three stamps. As the horse stamped the fourth time, the human observers would start to become alert. At the fifth stamp, they would very subtly relay their approval and prepare to clap. The horse, it turned out, was just very good at reading the body language of the people. Before long, the word got out about what was happening. It was not necessarily a hoax on the part of the horse's owner but just bias by the observers. Impressionable audiences continue to draw the same erroneous conclusions as did those who saw Clever Hans.[16]

It gets better. You would think that by the twenty-first century we would watch out for things like that. But I saw a television show in which people could demonstrate how intelligent their pets were. One family claimed that their dog could not only count but could also *read numerals on wooden blocks*. I swear I am not making this up. You can't make something like this up. The dog was on one side of a curtain. On the other side, the owner would pick up a block with, say, the numeral 7 on it and show it to the eager audience, then place it back down among the other blocks. When the curtain was raised, the owner would tell the dog to find the block with "7" on it. The dog would go right to it. The audience was enraptured, and the judges at least acted as if they really believed the dog could read numerals. I know you are ahead of me on this. The dog was not reading numerals; it was just sniffing the block that had the most recent human scent on it. In fact, the dog would start to go find the numeral 7 block *before* the person told it which number to find! You'd think that would have made the observers a little suspicious.

BIASED BRAINS

Not only does the human brain inevitably create bias, but the human brain also seems almost rigged to deceive itself.[17] Our minds are constantly recreating our memories. A memory is not a set of unchanging original information that our minds can consult anew each time. Instead, each time we think of something, our brains alter the memory and may introduce bias when it does so, then stores the modified version. We literally cannot remember what something was like before the bias. This is why it is so easy for law enforcement and national security investigators to instill false memories in people, who may then claim to remember something that did not, in fact, occur. Once the false memory is stored away, it is indistinguishable from a true memory. It is to avoid such bias that scientists keep lab notebooks, in which they write their results on the very day that they got them, rather than afterward, and why eyewitness testimony, which can be unintentionally delusional, must be checked against physical evidence such as crime scene DNA.

SLAVERY: A HISTORICALLY IMPORTANT EXAMPLE OF BIAS

One bias that prevailed in human history until less than a century ago is slavery. The owners simply defined slaves as not being human. Romans defined their slaves as "talking tools." While Thomas Jefferson, in the Continental Congress prior to the Revolutionary War, insisted that slaves were people, his fellow Southern delegates defined them as property. Often, defenders of slavery believed that slaves were happy with their lot.

Captain Robert FitzRoy, on whose ship the young Charles Darwin sailed around the world, believed that slaves were happy. Darwin disagreed with him. Since Darwin was a civilian passenger, he could disagree with FitzRoy without consequences. FitzRoy wanted to prove to Darwin that slaves were happy. So, during a stop in Brazil, FitzRoy had a slave owner bring out his slaves to meet Darwin. In the presence of the owner and the captain, the slaves told Darwin that they were, in fact, perfectly contented. Darwin pointed out something that would be obvious to any of us today: the slaves would not dare to complain in the presence of their owner, who would then punish them. When Darwin told FitzRoy this evidence was worthless, the captain's response was less than charitable.[18] Unfortunately, as in statements by prominent political commentator Bill O'Reilly indicate, this bias persists.[19]

RELIGION AND BIAS

One of the most important ways in which scientific thinking differs from other modes of human thought such as religion is that religion raises bias to the level of a virtue. The common practice during church services (and not just fundamentalist ones) is to get the worshippers to bow their heads, close their eyes, and empty their minds of the faculty of critical thought. Let the preacher's words flow into the brain without questioning them, thus bypassing the instinct of common sense. Religions use psychological techniques that are illegal for any corporation or government agency to use in a free society.

Religion is in particular the playground of confirmation bias. Millions of religious people (once again, not just fundamentalists) pray for someone to be healed from a sickness or rescued from a calamity, a kind of prayer known as *intercessory prayer*, in which the worshipper intercedes with God on behalf of the unfortunate person. When the person's health or luck gets better, the people who prayed notice it and use this piece of information to confirm their religious bias. They do not notice all of the times that the person for whom they prayed does *not* get better. People who receive prayer may get better more often than people who do not. What might actually be happening is a placebo effect: a sick person feels better, and maybe actually gets better, with the knowledge that somebody is praying for her, even if there is no divine intervention. What was needed was a double-blind study of the efficacy of intercessory prayer. And this is exactly what Herbert Benson and colleagues provided in a study published in 2006—famous as much for the fact that it was done and how it was done as for its results.[20]

The Benson study did not attempt to test any "God" hypothesis— that is, regarding the existence of God. But it did test the hypothesis that intercessory prayer enhances healing. Their specific hypothesis was that coronary bypass patients who received supplementary intercessory prayer would survive better than patients who did not receive supplementary prayer during a thirty-day period following surgery. Notice that I said "supplementary prayer." It would be theoretically ideal to have a group of patients receiving prayer from their churches, friends, and families, and another group (the controls) who did not. But this is impossible. You cannot prevent somebody from being prayed for. In this study, the patients in the experimental group received supplementary prayer and the patients in the control group did not.

And the hypothesis really is a hypothesis, not just a claim. Either the experimental group has statistically fewer deaths than the control group, or it does not. The hypothesis is falsifiable.

The sample of participants also had a fair amount of external validity, another concept I explained in a previous chapter: while they tended to be in their sixties, white, and male, they were diverse in their religious backgrounds and their geographical origin. Hospitals in Massachusetts, Minnesota, and Oklahoma participated. Most importantly, the experimental group and the control group did not differ initially in their demographic characteristics. Furthermore, the sample size was good: about eighteen hundred patients chose to join the study. Perhaps most interesting of all, the experimental and control groups did not differ in the number of people who indicated, in a survey, a belief in the power of prayer.

The group of researchers was impressive. The leader was Herbert Benson, who was not only an expert cardiologist but also had for decades studied the effects of the mind upon the body. I ran across a book of his from the 1970s called *The Relaxation Response*, which was about the therapeutic effects of relaxation techniques. Benson founded the Mind/Body Medical Institute at Massachusetts General Hospital. He has had a lifelong professional interest in the effects of things such as meditation and prayer upon recovery from illness. His collaborators included not only doctors and nurses but also experts in public health and even theologians.

To avoid the placebo effect, Benson divided the experimental group up into two subgroups: some of these patients *knew* that they were receiving extra prayer, and some of them knew only that they *may or may not* be receiving extra prayer. Benson's study removed an important element of participant bias.

There was another brilliant detail of experimental design. Remember I mentioned the double-blind arrangement to prevent the placebo effect. It was important in this study that the doctors and nurses who actually attended to the patients not know which group they were in. Patients received only sealed envelopes telling them which group they were in. The doctors and nurses did not know which patients were receiving supplementary prayer. Only the patients knew, as well as Benson and his team who never saw the patients.

Here was the perfect time for God to show himself, as religious people would expect.

Here are the results. The patients in the control group and in the experimental subgroup who did not know whether they received prayer had almost identical recovery rates. Intercessory prayer, itself, did not change the outcome. This constitutes experimental confirmation, though not proof, that intercessory prayer has no discernible effects on recovery from heart surgery.

What about the patients who *knew* they were receiving extra prayer? You probably expect that they had better recovery. But the exact opposite was true. *More patients died when they knew they were being prayed for.*

The explanation that seemed most plausible to me was the "performance anxiety effect." The patients who knew they were being prayed for may have suspected that they were part of a study in which God was being put to the test. They did not want to let God down. They were nervous, and this put stress on their newly mended hearts.

But another possible explanation is one that I have started calling the "potato chip effect." One of our graduate students (thanks, Tamara) helped me think of this one. The patients who knew they were receiving extra prayer may have thought, *Well, I don't have anything to worry about; God will take care of me*, and so they went back to the unhealthy habits (such as eating potato chips) that had gotten them into trouble in the first place.

This experiment has received criticism, of course, but mostly from religious people who do not like the way it turned out. Fundamentalists, for example, claim that the people who were providing the supplementary prayer were not real Bible Christians. The prayer was provided by an interfaith group and by two Catholic monasteries. Fundamentalists think it is highly unlikely that God would pay any attention to prayers from people like that. To their credit, Benson and associates tried to recruit some conservative religious groups but found none who were willing to participate.

Other fundamentalists reject the whole idea of testing intercessory prayer. Is it acceptable, from a religious point of view, to subject God to an experimental test? Well, Benson and associates subjected prayer, not God, to a test. But if you really want to know what the Bible says about this, you won't get a straight answer. In the Old Testament, Gideon and Elijah both put God to the test. Gideon put out a fleece on the ground to test whether God was actually speaking to him. Elijah actually made a public display of God lighting a sacrificial altar. But in the New Testament, Jesus says you should not put the Lord to the test. To these fundamentalists, God did not answer the prayers, because he was offended that someone would put him to the test, so he just let the people die whom he might otherwise have healed—to punish the patients for the sins of the blasphemous researchers.

SCIENCE AND DEMOCRACY

Thomas Jefferson, like other scholars of the Enlightenment, assumed that people would make rational choices when they had the opportunity to do so. Democracy is based on this assumption. And this opportunity included not just a free and secret vote but also the education to understand the issues. There is a reason that Jefferson both wrote the Declaration of Independence and founded the University of Virginia. I believe he would have been very upset to see how politicians and marketers create bias in our minds and use it as a harness to pull us in the direction that the rich and powerful want us to go. I often wonder if this process— government of and by people who are deliberately misled—will destroy democracy. I wish we had someone like Jefferson now who could help us figure this out.

We scientists deserve the esteem that we have for being unbiased— not because we *are* unbiased but because we, more than anyone else, recognize and, to the extent humanly possible, control for our biases.

DO IT YOURSELF!

Try to think of a personally important example of confirmation bias and see if you can think of a way to reduce its influence on your decisions.

CHAPTER 14

TRUST US, WE'RE SCIENTISTS

Almost all the time, you can trust scientists. But sometimes you cannot. This is particularly the case with pseudoscience, which can be confused with real science.

Pseudoscience is any set of beliefs that rejects the scientific method but which pretends to be scientific and thus basks in the respect that scientists usually receive for their work. In some cases, pseudoscientific organizations actually have a few scientists on their staffs.

While there are many kinds of pseudoscience, they tend to have the following characteristics:

- *Absence of self-correction.* Scientists continually seek new information in an ongoing test of each hypothesis. Pseudoscience, in contrast, staunchly defends its original tenets not only from internal review but also from external criticism.
- *Reversed burden of proof.* When scientists present a hypothesis, they consider it necessary to also present the evidence that tests it. Pseudoscientists, on the other hand, announce their hypothesis and insist that *you* must believe it unless *you* can prove that it is wrong. A scientist can have immense amounts of data, but any good old boy can come along and dismiss them—your data are not good enough to convince *him*. One of the ways that you can recognize pseudoscientific organizations is that they do not have their own laboratories in which to conduct their own research. They spend their time simply scoffing at the data that real scientists gather.
- *Overreliance on anecdotal data.* An anecdote is a story, often a single observation. Scientists use sets of data, which must, as explained

in a later chapter, be extensive in order to be considered valid. Religious beliefs consist almost exclusively of anecdotal data. Somebody prays, another person recovers from an illness, and to the religious person this proves that prayer causes healing. Pseudoscience does this also while pretending to be scientific.

- *Undefined terms.* Scientists define their terms as precisely as possible and risk criticism from other scientists if they do not do so. By leaving their terms vague, pseudoscientists can make their hypotheses unfalsifiable.
- *Refusal to pay attention to evidence.* Even if pseudoscientists receive the evidence for which they have asked, they ignore it.
- *Primacy of a political agenda.* All scientists, like everybody else, have political motivations. Real scientists try to keep them from invalidating their research; pseudoscientists use selected bits of scientific data to advance their political agenda.
- *Ridicule of or threats against critics.* Science, as a process, does not ridicule or threaten. Scientists, being human, sometimes do, but they usually receive criticism from other scientists when they do. Old-time religion relies heavily on threats: if you doubt the assertions of a certain preacher, he simply tells you that you are going to hell. For pseudoscientists, threats and ridicule are a routine practice.
- *Follow the money.* Pseudoscientists are often lavishly funded by corporations or political parties. The whole purpose of many prominent pseudoscientific "think tanks" is to confuse people about the scientific facts, thus disabling any opposition that citizens might take against the dangerous activities of the corporations and political parties.[1]

GLOBAL WARMING DENIALISM

Throughout the book, I have used global warming as an example of the challenges faced by scientists in getting valid data and testing it reliably. Now I wish to use global warming denialism as an example of pseudoscience.

Refusal to pay attention to evidence. One of my fields of research is climate science. I study the effects of warmer temperatures, which cause earlier spring budburst in deciduous trees. While there are many scientists more qualified to speak to global warming issues than myself, not too many of them live as I do in rural Oklahoma. This makes me something of a local expert. One time I was asked to give a presentation to a church group about global warming. Unusually for a small church in an Oklahoma town whose economy was almost completely based on oil, this group welcomed my message. Except, that is, for one man. He raised his hand and asked what I thought was a very good question. He explained that global warming had occurred extensively in the past, for example back in the dinosaur days. "The dinosaurs loved it," he said. So, what do we have to worry about today? The brief answer is that global warming in dinosaur days occurred much more slowly than it is occurring today; were modern global warming occurring more slowly now, it would not be a problem. But as soon as I said it was a good question and started answering it, he got up and walked away, and was out the door before I finished. I assume he was a "plant" from one of the oil companies and came to make his statement and leave. Walking out on the answer to your question is a literal example of refusing to listen to evidence.

Primacy of a political agenda. Another characteristic of pseudoscience, such as global warming denialism, is that it almost always is driven by a political agenda. All people, and all groups of people, have political views. But to pseudoscientists, the political position is all that really matters, and the science is just window-dressing. One of the leading denialist scholars, let's call him Dr. Lorax, champions the idea that carbon dioxide will cause trees and other plants to grow more, and these trees will absorb the extra carbon dioxide that humans are putting into the air. That is, plants are giant carbon eaters that will save the world. This cannot be true, because if trees were going to save the world, they would have already started. Plants are not now preventing the rise of atmospheric carbon dioxide levels.[2] Tropical rain forests are famous for absorbing large amounts of carbon dioxide from the air and transforming it into luxurious growth, but these forests have been so degraded by human activity that they are now net producers of carbon dioxide.[3] At the present time, temperate forests of North America, Europe, and Asia absorb more carbon dioxide than they release, but detailed studies predict that they will be net producers of carbon dioxide before the end of this century.[4]

But suppose that Dr. Lorax is correct. You would think that denialists would happily support global efforts to save the remaining intact forests and to replant the ones that have been destroyed. You would think they would be supportive of ideas such as the "cash for carbon" plan that successfully encouraged tropical farmers to leave rain forests intact in return for payments. A team of scientists, led by an economist, experimentally confirmed that the monetary benefits that the intact forests provided to Uganda far exceeded the money given to the farmers.[5] You would think that the denialists would be happy to support this, but they pretty much ignore it.

The pseudoscientists are not always on the Right end of the political spectrum. Many people who oppose vaccination are on the political Left. They base their opposition to vaccination on rumors that can be originally traced to an article that has been retracted and which was shown to be fraudulent.[6] They have interrupted vaccinations to such an extent that many diseases, once nearly eradicated, are starting to make a comeback.[7] Many medical researchers have experienced firsthand the hostility of the "anti-vaxxers," whose unfalsifiable beliefs persist even when their basis has been shown to be wrong. Unscientific attacks on genetic engineering also tend to come from the political Left.

Ridicule of or threats against critics. Denialists also respond not just to criticisms but even to sincere questions with ridicule. I wrote to one of the leaders of a large denialist organization and asked him in what ways the calculations published by climate scientists failed to prove the hypothesis that humans were causing global warming. (This was not Dr. Lorax, who was invariably polite to me. Instead let us call him Dr. Smoke, since his organization is also one of the leading opponents of tobacco taxes and smoking restrictions.) Dr. Smoke's response to me was brief and vivid ridicule. And why not? Dr. Smoke can ridicule scientists, so long as he is friendly to policy makers and to his fossil fuel industry donors. At least I have not, like climate scientist Michael Mann, received any death threats.[8]

Sometimes pseudoscientists resort to legal threats. About twenty-five years ago, I wrote a review of a book in which the author claimed to have invented a whole new theory of evolution. In my review I pointed out his errors. People like him spend a lot of time reading everything that anyone says about them, and he found my review, obscure though it was. He wrote to me (this was before extensive emails), saying that he had considered suing me but had graciously decided not to do so. He continues his lonely rants, this time unnoticed by publishers and reviewers. You can see why I have chosen to not identify him.

Follow the money. While some pseudoscientists, such as the geocentrists who claim that the whole universe revolves around the earth, are poorly funded, others are rolling in money. The denialists are a perfect example. They receive a lot of money from coal and oil corporations, which most certainly do not want citizens to use less coal and oil. They usually keep their funding sources secret. One such organization claims that its funding comes equally from corporations, organizations, and individuals. But remember if most of their money comes from a few oil-rich billionaires, those billionaires have corporations, they fund organizations, and they are individuals. Coal and oil can provide a lot more money than can agencies that fund global warming research, that is, for those climate scientists who receive such funding. I receive no funding for my climate science research.

Pseudoscientists often see themselves as victorious. When they find that nobody has come forward to absolutely prove them wrong, they interpret this as proof that they are correct while, in reality, it may be that nobody wants to subject herself to the barrage of verbal attacks and legal threats that she would receive for doing so.

CONFLICT OF INTEREST

Pseudoscientists like to think that, because they are not actually lying, their opinions are not influenced by the funding they receive. But they would be superhuman in their virtue if this were so. All people, including true scientists and everybody else, hesitate to bite the hand that feeds them. As Upton Sinclair said, "It is difficult to get a man to understand something when his salary depends upon his not understanding it."[9]

Besides being a good example of pseudoscience, global warming denialism is a very good example of conflict of interest. The over-whelming consensus of scientists and the majority view of most people, including Americans, is that we should burn less coal and petroleum. But major energy corporations do not want us to use less coal and petroleum. Even though, in any possible scenario of future energy sources, we will continue to burn a lot of coal and petroleum, the corporations that provide it do not want us to burn less, even if we still burn a lot.

One of the ways they do this is to provide, directly or through private foundations, a lot of campaign money to legislators who make the rules about the use of energy. Legislators have never been unbiased, and they have even more reason to be biased on this subject. Oklahoma senator James Inhofe received large amounts of campaign money from oil companies.[10] Whether it was intentional on his part or not, he proceeded to make patently ridiculous statements about global warming. Senator Inhofe said that global warming cannot occur, because the Bible says God will not allow it to occur. The biblical verse he cited does not, in fact, say what he claimed it said. He gambled and won: most of his followers did not bother to look up the verse and read it for themselves. Inhofe also threw a snowball during a Senate session to prove that, since winter is still cold enough for snow, then global warming cannot be occurring.[11] If it ever snows anywhere anytime, he seems to imply, then there is no global warming.

One of the best examples of pseudoscientific conflicts of interest involves pesticides. In the two decades after World War II, massive amounts of pesticide were sprayed from airplanes all over the American landscape in an attempt to control agricultural pests and insect disease vectors. A biologist, Rachel Carson, pointed out that these pesticides, while perhaps harmless in their environmental concentrations, became more concentrated in the tissues of animals in the food chain, especially at the top of the chain.[12] Moreover, insect pest populations evolved quickly and became resistant to the pesticides. She explained that using less pesticide, and using it in a more targeted fashion rather than spraying it from airplanes, was safer for the environment *and* more effective at controlling pests. She assembled an impressive set of evidence and explained it in language that anyone could understand in her 1962 book *Silent Spring*.[13] (That is, the songbirds would die because of the pesticides they ate in their prey, rather than from direct exposure to spray.) It remains one of the most important books in history. Carson died in 1964.

When Carson's book came out, the chemical corporations were not pleased. These corporations made money not just from selling pesticides but also from selling *lots and lots* of pesticides. If governments and farmers used small amounts of pesticides rather than large amounts, the corporations would lose a lot of money. The corporations accused Carson of saying that nobody should ever use pesticides. One recent book, whose author had apparently not read *Silent Spring*, described it as "an angry, raging, no-holds-barred polemic against pesticides."[14] This is not what Carson said. In chapter 10 of *Silent Spring*, Carson condemned the spraying of pesticides "indiscriminately from the skies," for instance to control fire ants, and said instead that it was cheaper and more effective to spray the ant mounds directly.[15] She defended the limited use of pesticides. I believe that, had she lived, she would have approved the use of the pesticide DDT to spray around doors and windows and on bed nets to prevent mosquitoes from spreading malaria to sleeping people where it is prevalent in Africa. This is, in fact, now being done. This practice is safe for the people and the environment.

The same author who called Carson's book an angry polemic also claimed that it was "short on data and long on anecdotes," for example simply taking the word of a retired bird-watcher rather than using scientific data. If this author had actually looked at Carson's book, however, he might have noticed that she cited *754 references* at the end, most of them from official government reports.

But some spokespeople for the chemical industries still say that Carson singlehandedly convinced the world to stop spraying mosquitoes, which led to a surge in mosquito populations and the deaths of thousands of people from malaria. This is simply a lie. Carson convinced the US government to ban the agricultural use of DDT but not its use for insect vector control—and only in the United States. One would think that, in the years since Carson's death, the lies about her would have stopped.[16] But anti-environmentalist websites and books continue to propagate this misinformation. Oklahoma senator Tom Coburn blocked a Senate resolution of respect for Rachel Carson in

2007, the one hundredth anniversary of Carson's birth, citing this incorrect information.[17]

Another way in which Carson's story is an example of bias is that the chemical corporations said that she was hysterical because she was a woman.[18] As incredible as this may seem today, it was a widespread bias at the time: you can't trust what women say because they get hysterical. The Latin origin of the word *hysterical* itself refers to female reproductive organs. Carson was held to a much, much higher standard of evidence as a result of this bias. Fortunately, her evidence was so good and so clearly presented that her viewpoint prevailed.

Examples continue to emerge of scientists working for large chemical companies, minimizing the potentially major health risks of the compounds that they study. Some scientists who work for Monsanto claim that the corporation did not want them to find that glyphosate, one of their most important herbicides, may cause cancer.[19]

JUMPING TO CONCLUSIONS

Sometimes scientists can do things during their research that, while not deliberately dishonest, might slant their results toward having more credibility than they deserve. We can *unconsciously* "cook our data" to look better than they actually are. In the competitive world of research funding and scientific jobs, the temptation can be very strong to do this. Let me give a couple of examples.

First, consider which data to keep and which to throw out. I alluded earlier to statistical outliers and how to eliminate them from a data set. We scientists, all the time, throw out data that are somehow faulty for reasons that have nothing to do with the experiment. An experimental plant or animal might get sick for some reason unconnected with the experiment, and it would be misleading to include that plant or animal in the data that we analyze. But when we throw out these data, we do two things. First, we admit it. Second, we try to explain why. Unfortunately, in many cases of high-stakes scientific research, this is not done.

Sometimes, if a scientist finds herself having to throw out a lot of data, this might be an indication that something is wrong with the experiment other than just a little bad luck. In one of our experiments with lab rats, we found that a lot of them were just not very healthy. Rather than throw out a big chunk of data, we had to figure out what was going on. As it turned out, the rats were having an allergic reaction to the shaved pine bedding in their cages. When we switched to aspen bedding, the problem went away.

But sometimes it can be very tempting to throw out inconvenient data. I said previously that scientists accept the 5 percent significance level for accepting results. Suppose that the calculation comes out 6 percent instead of 5 percent. And suppose further that throwing out just one data point would improve the calculation to 5 percent. There is a strong temptation to find some rationalization to throw out that one data point. I do not believe I have ever done this, but I have a deceptive subconscious mind just as you do. The best thing is to have a big enough sample size, or to repeat the experiment enough times, that one data point does not matter so much.

Second, consider pseudoreplication.[20] Pseudoreplication is where you, in effect, measure the same data point over and over, rather than measuring truly independent data. Obviously, it is dishonest to (for example in a toxicity study) measure the same organism over and over again. To do so is not to increase your replication but to inflate it spuriously, hence the name. But sometimes you can commit pseudoreplication without intending to, for example as a result of trap-happy animals (see chapter 13).

One recent example of a scientist jumping hastily to a conclusion was the announcement that arsenic-based life-forms had been discovered, right here on Earth.

Life is based on DNA, which itself is built on a scaffold that includes phosphorus atoms. Phosphorus atoms are chemically similar to arsenic atoms. Although arsenic is poisonous to all known life-forms on Earth, it is possible that somewhere—on Earth or on another planet—a life-form could have evolved that used arsenic atoms instead of phosphorus atoms in its equivalent of DNA. And a brilliant young scientist named Felisa Wolfe-Simon claimed to have discovered just such a life-form in Mono Lake in eastern California, a lake whose waters contained high, though not deadly, amounts of arsenic.[21] NASA, which had sponsored her research, was excited also because if such a different kind of life could evolve right here on Earth, why not on Mars? So, let's go to Mars and find it! For a brief while, Wolfe-Simon was a rock star scientist.

It turns out that the bacteria discovered by Wolfe-Simon were probably not using arsenic to make their molecules but were merely sequestering it in such a way as to keep it from interfering with cellular processes.[22] Admittedly, it is hard to tell the difference between arsenic-based "DNA" contaminated by phosphorus and phosphorus-based DNA contaminated by arsenic.

DOWNRIGHT DISHONESTY

Scientists are among the most trustworthy and honest people in the world. It is true that scientists, on the average, are more honest and trust-worthy than, for example, politicians, preachers, and business leaders. This is because the scientific standard of success is, or should be, truth. There are honest politicians (we call them statesmen), honest business-people, and honest preachers. But they appear to be rare enough that we lionize them in literature and the media. *Mr. Smith Goes to Washington* is an example of one honest politician in an ocean of corruption. We also love to hate the dishonest ones, the way Sinclair Lewis hated Elmer Gantry and the way Mark Twain hated the dishonest businessman in *The Man That Corrupted Hadleyburg.*

But, although dishonesty is comparatively rare in science, there are numerous examples of it. I will tell a few classic and modern stories.[23]

Perhaps the most famous hoax in the history of science was Piltdown Man. In 1906, an amateur fossil hunter named Charles Dawson was digging around in a gravel quarry near Piltdown in England. He claimed to have found, in this quarry, bones of prehistoric humans that were "the missing link" between humans and apes. The bones came from a skull that had apelike teeth but a relatively large brain. This sup-posedly proved two things: first, that the initial stage in the evolution of humans from apes was the growth of the brain; second, this stage occurred in what is today England. Other "missing links" had been dis-covered, for example "Java Man" in Indonesia. But Dawson insisted that such fossils as Java Man were not human ancestors. They were in the wrong part of the world; everyone knows that intelligence evolved first among the earliest Europeans who remained, Dawson believed, the most intelligent people in the world. Maybe in Indonesia there were some apes evolving the ability to walk on two legs, but in England they were evolving intelligence.

British fossil experts like Arthur Smith Woodward were so eager to believe Dawson that they overlooked problems that should have been obvious. Among these problems were that the Piltdown Man skull had been filed to make it look like what Dawson wanted it to look like, and it had been stained to look older than it was. It looked like a human upper skull with an ape jaw. And, in fact, it was. Although recent evidence implicates at least the involvement of a minor curator at the British Museum, most historians have believed that Dawson perpetrated the hoax himself, since he was the principal public defender of it and basked in the fame associated with it. Some historians speculate that Dawson would eventually have told everyone what he had done—that he just wanted to play a practical joke—but he got killed in World War I before he could fess up.

Another famous hoax was perpetrated by a British educational psychologist, Sir Cyril Burt. He made measurements that supposedly proved that white people were much more intelligent than people with darker skin. After Sir Cyril's death, however, it turned out that his measurements may have been fabricated. He apparently not only just made up numbers, but he also claimed to have had a research assistant, Margaret Howard, who it turns out did not really exist.[24]

In the above cases, the frauds were motivated by the desire of some scientists to prove that the British were the smartest people in the world and had been so for a half million years. But in most cases of scientific fraud, the principal motivation seems to have been self-promotion of egos and careers. This is the most likely explanation of the fraud committed by not one but three separate Dutch social psychologists (Diederik Stapel, Dirk Smeesters, and Jens Förster) between 2011 and 2014.[25]

FOLLOW THE MONEY

But there is another obvious motivation: money. If you want to find scientific fraud, follow the money. This is why fraud is much more common in medical research than in more poorly funded ecological field research. And in medical research, one hot new field—and apparently a magnet for fraud—is the production of bio-identical embryonic stem cells.

Bio-identical embryonic stem cells would be a dream come true for medicine. Embryonic stem cells (unlike the adult stem cells in our bone marrow) have the flexibility to develop into any kind of cell in the body. They have to; they are the cells in the little clump of the very early stage embryo that, in fact, do become all the cells of the fetus and the infant. This includes cells, like nerve cells, that normally cannot make copies of themselves. Suppose your nerve tissue is damaged by Alzheimer's disease, Parkinson's disease, or physical injury. It will remain damaged for the rest of your life. But if you can inject some embryonic stem cells in the damaged nerve tissue, those cells might develop into new nerve cells and heal the injury. Now, the problem with embryonic stem cells is that they come from an embryo that is genetically different from you. Therefore if someone injects embryonic stem cells into your tissue, your immune system may attack them as foreign invaders. But if those stem cells have had their nuclei removed and *your* nuclei inserted instead (that is, if they are bio-identical), then your immune system will not attack them.

A South Korean scientist named Woo-Suk Hwang rocked the world in 2004 when he claimed to have produced embryonic stem cells in which the nuclei had been replaced.[26] It seemed that the medical availability of bio-identical stem cells was right around the corner. Then other scientists began to notice suspicious things. Hwang used some practices in his laboratory that were not quite ethical. This got scientists to look more closely at his results. And that is when they discovered that his genetically engineered stem cells were just Photoshop figments.[27] Scientists in Oregon appear to have legitimately accomplished for other primates what Hwang claimed to have done for humans.[28] But there was not as much celebration this time—it's hard to celebrate a legitimate breakthrough after a fraudster has made us all feel like idiots for believing his claims.

Medical researchers would very much like to find a way of turning adult stem cells, which unlike embryonic stem cells are readily available, into something as developmentally flexible as embryonic stem cells. In 2006, two scientific teams, one led by James Thomson (the original discoverer of embryonic stem cells) in Wisconsin and another by Shinya Yamanaka in Japan, figured out how to do genetic manipulations on adult stem cells to induce them to be almost as good as embryonic stem cells.[29] They had created "induced pluripotent cells." But whenever genetic manipulations are done, there is the possibility that something might go wrong. There must be a simpler way to induce adult stem cells to become pluripotent. In 2013, Haruko Obokata and other scientists in Japan claimed to have found that simpler way: just briefly expose some adult stem cells to mild acid, and they will become pluripotent.[30] But hardly a year had passed before it became apparent that their claims, even if not fully fraudulent, were premature and overblown. Here we go again, everyone said. In 2014, Obokata's mentor on this project, Yoshiki Sasai, hanged himself.[31]

Scientific societies, such as the American Association for the Advancement of Science, are taking these problems seriously.[32] Integrity includes not only to be honest but also to conduct research that is in the interest of humanity and has respect for human rights.

AN INDUSTRY OF FAKE PAPERS

In the old days, when scientific papers were all actually published on paper, there was fairly good quality control. It was expensive for journals to publish papers—a lot of ink, paper, printing equipment, and postage—and if any journal got a reputation of publishing untrustworthy papers, it would quickly go out of business. But since the rise of online publishing—something that was inevitable and can be really good—a whole new kind of problem has arisen: fake scientific papers. Today it costs hardly anything to set up a "scientific journal" and start issuing online papers. There must be dozens of such journals now on the web. If you pay them, they will publish almost anything. It is an analogous problem to the difference between blogs and books released by legitimate publishers. Anybody can write a blog and say anything, no matter how false, and there is no editor to say, "Your outrageous claims will tarnish our reputation and make us lose money or even go out of business!" No matter how crazy your blog is, it will stay in business, since it may cost you nothing. If you pay a provider to host a website, they make money from you, not from your readers. Websites and blogs are the new equivalent of subsidy publishing, where certain companies will print up whatever you pay them to. I like to think that my science blog is reliable,[33] but there is nothing about the blogging process itself that assures this to be the case.

A reporter for the *Ottawa Citizen* whipped up a fake paper and sent it off to some of these "peer-reviewed" journals.[34] He took half of the title and half of the material from geology, and the other half from hematology: the article "Acidity and Aridity: Soil Inorganic Carbon Storage Exhibits Complex Relationship with Low-pH Soils and Myeloablation Followed by Autologous PBSC Infusion." You've got to admit it was creative, especially with the author's invention of "seismic platelets." He got several acceptances. A few noticed his plagiarism but told him to do a little rewording and it could be published.

What's there to get upset about here? It is true that published papers are still the coin of the realm for academic success. But anyone who hires an applicant or awards a grant to someone whose publications have titles like this has only himself to blame. These "peer-reviewed" papers apparently have no peer review, depending on your interpretation of peer. One could interpret "peer" in such a way as to make all of us peers (i.e., micturators). Anyone who is hiring a scientist should give credence only to journals that are known to be reliable, even if not widely read. You cannot distinguish between real and bogus publishers by asking whether they charge a fee to the author; nearly all journals are nonprofit and have "page charges" to cover their costs.

And apparently the costs of scientific fraud, or at least scientific haste, can add up. From 1992 to 2002, $58 million of National Institutes of Health (NIH) funds were spent on research that was later retracted.[35] While this is a small part of the NIH budget during that time and is incredibly less than the financial scandals of many major corporations, a million here and a million there starts to add up to some real money.

Peer review (as noted in an earlier chapter) is frequently a capricious affair. But at least it can be pretty good at catching fakes. A few years ago, as a reviewer, I caught a plagiarized paper. However, once a paper has been published, it circulates not only among scientists but also among science writers and reporters. It is nearly impossible to correct the mistake once it has disseminated in the public sphere. In response to this problem, the American Association for the Advancement of Science, the publisher of *Science*, began an initiative in November 2017 in which reporters could contact leading scientists right away to check the accuracy of scientific information that they use in their stories.[36] Although this does not protect the reading public from invalid science that has not yet been exposed, it will at least help to reduce the amount of scientific misinformation that would otherwise circulate.

So, go ahead and have a laugh at the articles published by bogus journals, and if you believe them, the joke's on you. On a related note, I'm thinking about starting a science journal. I only charge $10,000 per article. If you are interested in publishing there, let me know, and remember I'm a peer.

DO IT YOURSELF!

Think of an information source (e.g., an organization) whose information might be unreliable.

CHAPTER 15

TRAPPED

The most fundamental human urge is not food, shelter, or even sex. The most fundamental urge of each human is to understand her world and to make sense out of her experiences. Each of us creates a theoretical model of the world—whether we are aware or not that we have done so—that defines our role and significance in the world and a general landscape of how we should live.

But once you have made such a mental model of the world, you are trapped. You feel the need to interpret everything that happens to you and everything you see in terms of that model. If your model contrasts sharply with reality, this becomes increasingly difficult to do. At some point, as the contradictions between your beliefs and the facts accumulate, you may eventually realize that your model needs to change.

Or not. When some people encounter facts or have experiences that contradict their beliefs, they will invent increasingly complex webs of explanation to keep their beliefs from collapsing. The process of constructing and maintaining these webs of delusion is called *confabulation*. Confabulation is not lying but producing a whole false mental world in which to live. Author Sam Kean calls it "honest lying."[1]

Since religion is such a fundamental instinct and addresses our most basic questions, it provides some of the best examples of confabulation. One example of this is Christian fundamentalist creationism. The purpose of this chapter is not to beat up on creationists but to use their beliefs as an example of how desperate confabulation can become.

There are many kinds of creationists, but I will use young-Earth creationists as an example. Young-Earth creationists are generally considered to be people whose entire set of beliefs is built on Genesis 1. But actually, their beliefs are built upon Genesis 1, 3, and 6:

- They believe Earth is young. God created it supernaturally only about six thousand years ago (Genesis 1).
- They believe Earth was created perfectly and that all imperfection was introduced at the time of the Curse, also known as the Fall of Man (Genesis 3).
- They believe a Flood covered Earth entirely at the time of Noah, killed all the animals except those on the Ark, and produced all (or almost all) of the layers of sedimentary rock about four thousand years ago (Genesis 6).

They are trapped by these three beliefs. Whenever they encounter a fact that contradicts any of these beliefs, they have to come up with an excuse. Once they have enough of these excuses, they produce a confabulated theory of the universe.

Consider the first belief. The red shift of light coming from distant galaxies makes those galaxies appear to be so far away that light itself had to travel through the near vacuum of space for billions of years to get to Earth. What do the creationists do with this inconvenient fact? They claim that God not only created all the galaxies but created the light between them and Earth, giving the light just the right amount of red shift and keeping it in perfect focus. This allows them to continue believing that the universe is only six thousand years old. Moreover, geologists can use radiometric dating to determine the age of a volcanic rock by measuring the amount of uranium and lead in crystals found in the rock. These measurements allow geologists to determine that the volcanic rocks—and therefore the sedimentary rocks between the volcanic layers—are millions or even billions of years old. What do the creationists do with this inconvenient fact? They claim that God manipulated the uranium and lead ratios in the crystals to make the rocks look old.

Consider the second belief. They claim that when God created Earth, it had no predators or parasites or diseases or death. But both predators and prey have many complex adaptations—such as camouflage, the speed with which predators fly or run, their keen eyesight, and the instincts that both predators and prey have—that make no sense in a Garden of Eden in which lions and *T. rex* munched only on plants and in which peregrine falcons ate seeds. Parasites have astonishingly complex life cycles. And if there were no diseases, why do animals have the physiological ability to fight them? The most genetically complex example is the vertebrate immune system. Finally, aging and death are programmed into our chromosomes; no set of conditions can stop the inevitable march of senescence. What do the creationists do with these inconvenient facts? They say that God created all of these structures and processes at the time that he chased Adam and Eve out of the Garden. This would have amounted to a nearly complete re-creation of the world at that time.

Finally, consider the third belief. If all of the fossils were produced during a single, worldwide Flood, why is it that they are found in an evolutionary order? The oldest, lowest rocks that have visible fossils contain no amphibians, reptiles, birds, or mammals—in fact, no terrestrial animals or plants at all. There are no dinosaur fossils in the youngest rock layers. The oldest forests contained no flowering plants. What do the creationists do with these inconvenient facts? Some creationists claim that the terrestrial animals swam like hell to get to higher ground and that's how they got buried in the uppermost layers of sediment. I suppose the flowering plants had to do this also? But the problem is that there are no exceptions to this evolutionary order. A century ago, British biologist J. B. S. Haldane issued a challenge: show him just one mammal, *just one*, maybe a rabbit, that got washed into the lower layers of rock, and he would reconsider their assertion.[2] Other creationists, unsatisfied with the belief that mere swimming could produce a perfect evolutionary order of animal fossils, claim that God scooched the organisms around during the Flood to produce the pseudo-evolutionary order.

The creationists have constructed an elaborate confabulation to accommodate the facts revealed by science. God created the red shift, manipulated the uranium and lead ratios in the crystals, recreated the whole world at the time of the Fall of Man, and moved the doomed plants and animals during the Flood to produce a fake evolutionary order. Do they have any evidence for any of these assertions? No. They do not even have biblical evidence, for the Bible makes none of these claims. They just made these things up.

Scientists must also fit observations into an explanatory framework.

But science does not confabulate. The reason is that each of the assertions regarding the relationship of the facts to the framework can be, and are, tested against the evidence. Creationists construct a house of straw, which cannot stand in the real world of stormy facts. They mentally insulate this house from external facts. But scientists construct a house of strong beams and bricks, testing each one of them against the facts before the storm comes, and then we allow the storm of external validation to break upon the house. Creationists are trapped by their framework, but scientists are not trapped by theirs. Creationists believe that to question the framework of their house is to question God himself, but scientists will let anyone tramp right into the house and look at the framework for herself.

DO IT YOURSELF!

Try to think of a group of people (e.g., a cult) whose beliefs have trapped them into confabulation. Then examine what they have tried to do about it.

CHAPTER 16

WHAT EXACTLY DO YOU MEAN? WHY SCIENTISTS (SHOULD) CAREFULLY DEFINE THEIR TERMS

One difference between scientific thinking and other modes of thought is the careful definition of terms. A scientist is someone who will get bent out of shape if you call a beetle a bug. What both you and the scientist need to realize is that, in everyday conversation, a "bug" is any small organism that is potentially harmful or at least irritating. Thus, in everyday speech, a bug can be any insect or even a germ. To a scientist, a bug is an insect in either the order Hemiptera or the order Homoptera. In scientific work it is important to not call a beetle (order Coleoptera) a bug, since the exact identity of the insect may be important to know. For example, in agriculture, the distinction may be important. If all you are going to do is spray pesticides and kill everything, you can call them all bugs. But if you are using biological control agents, such as beetles, to control aphids, which really are bugs, you cannot use the word "bug" loosely.

Some of the terms we use are very simple but represent very complex processes. One example is the word "eat." Do many bacteria eat sugar? I dared to say so in one book manuscript, and a reviewer wrote, "Bacteria do not eat! They metabolize!" Sorry. Ambrose Bierce, in *The Devil's Dictionary*, defined "eat" as "to perform successively (and successfully) the functions of mastication, humectation, and deglutition."[1] Let's leave it at that. "Digest" is another simple word that denotes a series of complex processes: breakdown of food chunks into large food molecules in the stomach, breakdown of large food molecules into small ones in the first part of the small intestine, and absorption of small food molecules into the blood by the small intestine. And more. Nobody learns what words mean by analyzing them; we simply grow up hearing adults use words like "eat" and "digest," and learning from context.

One interesting example of a term that everybody uses loosely, even scientists, is the word "adaptation." It is a word that, in everyday speech, means "coping" or "adjustment." It is such a common word that it was the title of the Nicholas Cage movie based on Susan Orlean's book *The Orchid Thief*. Most of us understand it in context. But scientists need to be more precise, especially with a word like this.

There are lots and lots of ways in which organisms can adjust to circumstances or to changes in circumstances. That is, there are many *processes* that can be called adaptation. Consider how an animal might make sure it gets sufficient oxygen to its cells. Its cells might experience low levels of oxygen availability for any of several reasons. The animal might have just exercised vigorously, and its cells have used more oxygen than the blood can replenish. Or the animal might live at a high elevation, where the air and therefore the oxygen are thin. Red blood cells carry most of the oxygen in the bloodstream. Doctors and nurses call them RBC, which simply means red blood cells. *RBC* is no more precise than *red blood cells*, but the doctors want you to think they are smart. (Speaking of precise terms, *elevation* refers to height above sea level while *altitude* refers to height above the ground. Now you know.)

There are at least two processes by which an individual person can "adapt" to low oxygen availability and get more oxygen to his or her cells. One process is for the heart and lungs to operate more vigorously, thus delivering more oxygen to the tissues. This is simple physiology. Well, maybe not so simple, since it requires the brain stem to monitor carbon dioxide levels in the blood plasma and send a signal out to the muscles of the lungs and heart. This is something your body does without having to alter its pattern of gene activation.

The other process is for the bone marrow to begin producing more red blood cells. This will allow each milliliter of blood to deliver more oxygen to the tissues. The person has to stay at a high elevation for about a week before his or her red blood cell count begins to increase. These are actually changes in gene activation and are often called *acclimatization*. This explains the popularity of athletic training at high elevations even for events that take place at low elevations.[2]

There is another process by which an individual person can "adapt" to low oxygen availability, but it is something the person must do from earliest childhood. People who grow up at very high elevations, such as in the Andes, actually develop bigger lungs relative to their body size.[3] A person cannot do this once he or she has already grown. Because this process implies that body structure can be molded into a different shape, like plastic, it is often called *plasticity*. Acclimatization can occur only within the constraints of body structure while plasticity can change the body structure to a certain extent.

Physiological adjustments, acclimatization, and plasticity are the only things an individual organism can do. The organism cannot change its genes. Only evolution can do that, by changing the relative proportions of different gene variants in a population over time. Natural or artificial selection (explained in the next chapter) results in evolutionary adaptation. Groups of people, such as people living high in the Andes and in the Himalayas, who have lived for many generations at high elevations, have actually evolved the ability to produce more red blood cells.[4]

Thus, there are four distinct processes that can be called adaptation. And the physical characteristics that result from each process can be called adaptations. This gives eight possible meanings to the word "adaptation." And there are even more.[5] Evolutionary scientists use "adaptation" only for the processes or products of evolution. In medical research, you can still read about the other kinds of adaptation, perhaps because relatively few medical researchers think about evolution.

And from there it gets complicated. Consider this example. Over evolutionary time, there are many ways that plants have adapted to hot, dry conditions. One of these ways is a kind of metabolism known, for reasons I will not explain, as "C_4 metabolism." There may be as many as 7,500 species of plants, mostly grasses of hot, dry habitats, that have this kind of metabolism. And this adaptation appears to have evolved from separate evolutionary ancestors as many as forty-five times.[6] So, how do you quantify this adaptation? C_4 metabolism is *an* adaptation, but it originated by evolution forty-five times, and evolution has produced 7,500 species that use it. It is perhaps best to say that C_4 is an adaptation that evolved forty-five times. It would not be correct to say that this adaptation occurred 7,500 times. This is what Joseph Felsenstein called the *phylogenetic effect*.[7] You count the number of evolutionary origins of the adaptation, not the number of species that have it.

Another reason that scientists use words very precisely is that some other scientists are just waiting for the chance to shoot them down for the slightest lapse of precision. Sometimes the hypercritical scientists will say, "Does the learned doctor mean to imply. . ." and then give an outlandishly incorrect interpretation of the phrase in question. Actually, it does not mean that the author meant to imply, but that the critic meant to infer—or "misconscrew"—something ridiculous. Scientists write and speak precisely to keep such trivial disputes from even getting started.

Scientists pay close attention to physiological processes and to evolution. They cannot afford to be imprecise in their use of words like "bug" or "adaptation." Scientists are so careful with wording and are so aware of the tentative nature of scientific conclusions that nonscientists may think they are less confident of things than other people. Actually, scientists are *more* confident because they know the limits of precision of their beliefs, as described in an earlier chapter. This perceived lack of confidence is something that pseudoscientists exploit by saying that "scientists aren't sure." But if my statistical analysis says that $p = 0.015$, I can say that I am 99.5 percent confident that I am right, which is more than preachers and politicians can say, even if I have lots of caution words (such as "it appears that . . .") in my scientific statements.

Fortunately, it *is* possible to write precisely, clearly, and beautifully all at the same time. Science communication is a science and an art.

DO IT YOURSELF!

Think of a word that you frequently use but that could be misunderstood by somebody.

BIG IDEAS

Philosophers and theologians think they have the Big Questions arena to themselves, especially when it comes to human nature. However, science not only addresses the Big Questions but also has for centuries been conquering BQ territory from which philosophy and religion have had to retreat.

CHAPTER 17

NATURAL SELECTION: THE BIGGEST IDEA EVER

What is the most important idea in science? For me, it is an easy choice. The most important scientific insight is the discovery of *natural selection* by Charles Darwin. I did not say the discovery of *evolution*; he assembled compelling evidence for this also. But natural selection was his, and science's, greatest discovery. Natural selection explains *how* evolution works.

And not, as I will explain, just the evolution of organisms. But the evolution of *everything*.

There are two realities about the world and worlds that we all know. One is that there is an overarching order to it. We could call this the laws of nature. And these laws operate everywhere, which is why we call everything we know the "universe," from the Latin for "one." We now know that these laws are a part of nature, rather than something imposed on nature by the continuing will or whim of deities. The sun is not Hyperion's chariot, nor the moon an "orbéd maiden with white fire laden," as the poet Shelley called it.[1] The universe is not, as the ancients thought, a chaotic ocean ("the waters which were above the firmament" in Genesis[2]). It is an orderly cosmos. This part of our understanding of the cosmos was forced on the Western mind by people such as Galileo and Newton. Simple equations of gravitation and momentum explained the behavior of all the planets and their moons.

But we also know that this cosmos still contains a lot of chaos. There are lots of things that seem, Newton or no Newton, to just happen for no apparent reason, usually in the quotidian realm of our daily lives. For all its peaceable order, the universe still contains a lot of terror.

Natural selection ties these two aspects of reality together and does so by the operation of natural law. Darwin's insight was so devastatingly simple that many of us have had the same reaction that Darwin's friend Thomas Henry Huxley had: how extremely stupid not to have thought that![3]

EVOLUTION FOR CAT LOVERS

Here's how natural selection works. We will, like Darwin, apply it to organisms; let us use many people's favorite animal, the cat. Bartholo-Meow, again.

First, it is obvious that there is no such thing as *the* cat. Cats are all one species (*Felis catus*), but there is a tremendous amount of diversity in body size and shape and in how long their hair is. There are long-haired cats that choke on their hair balls, and nearly hairless cats that shiver but are preferred by allergic cat owners, and everything in between. To Darwin, there was no archetypal "cat" but rather a range of variation that makes up the cat species, as is true of any species.[4] And, as far as we can tell, this variation is random. No telling when a new genetic muta-tion will come along or what it will be. There is, in the cat species, a whole range of hair lengths.

Second, in a cold environment, long-haired cats would survive better than the short-haired ones. The long-haired cats would, therefore, have more kittens and pass on whatever genes made them long-haired; the next generation of cats would have relatively more longhairs than the previous generation.

Third . . . there is no third. That's it. The population of cats has evolved longer hair. That's natural selection.

And that's what *evolutionary fitness* is. Fitness is simply successful long-term reproduction. In this scenario, the longhairs were more fit. A muscular animal that leaves no offspring may be "physically fit" but his evolutionary fitness is zero.

Notice what did *not* happen. The cats did not all develop longer hair. The individual cats did not change. Longhairs have long hair, and shorthairs have short hair, and that's that. What happened is that the longhairs had more kittens, and the shorthairs fewer. Evolution cannot happen in one generation; instead, from one generation to the next, the longhairs became more numerous in the population.

Another thing did not happen: natural selection does not require that any of the cats actually die, except in the normal course of their lives. The more successful cats left more offspring; the less successful cats left fewer. It is possible that some of the kittens died. But there was no bloody battle with the "fittest" killing the less fit.

So don't let anyone tell you that evolution is random. The genetic variation upon which natural selection acts is, to all appearances, random, but natural selection is not: in this case, it selected the cats that were most consistent with the laws of nature—the ones whose hair held in their body heat the best. You could wait forever for the cat population to, by sheer chance, evolve longer hair. But natural selection imposes a directionality on evolution.

There is one other random element to evolution, though. The fact that the environment got colder was, to us, and certainly to the cats, unpredictable. You can think of it as a random event. The environment might just as easily have gotten warmer, in which case the shorthairs would have been fitter and evolution would have taken a different direction.

Darwin was quite aware that he was bringing the mysterious process of evolution into the realm of natural law, just as Newton had brought the motion of things on Earth and in heaven into the realm of natural law. In the last paragraph of *The Origin of Species*, perhaps the modern world's most famous and influential book,[5] Darwin wrote, "Whilst this planet has gone cycling on according to the fixed law of gravity."[6] In 1784, philosopher Immanuel Kant wrote that there would never be "a Newton for the blade of grass."[7] But Darwin had become that very thing. In doing so, however, he did not make the course of evolution lead to a totally predictable outcome. Instead, he brought together the world of chaotic, random genetic variation and environmental changes with the orderly process of natural selection.

It was a whole new way of thinking. No longer was there fate, in which everything that happened was foreordained—unless you actually knew the energy and mass of every particle and atom in the universe, you could never predict what the future was going to be. Nor was there just chaos.

In order to understand natural selection, Darwin extensively studied *artificial selection*, for example crop and livestock breeding. It is actually the same process, though the rules of success are different from those of natural selection. In artificial selection, the breeder determines which plants or animals to propagate for the next generation. Instead of corresponding with only fellow scientists, Darwin interacted with crop and livestock breeders, especially pigeon breeders, to get the straight information from them about this process.[8]

By studying artificial selection, Darwin learned two very important things about natural selection. First, wild species contain a vast amount of genetic (he called it heritable) variation that is invisible to our eyes but which can be distilled, as it were, out of their populations by breeders. All the weird characteristics of pigeon breeds, such as fancy feathers, were present but hidden in wild pigeons. Second, he found that selection can occur rapidly if it is strong and consistent. Artificial selection is strong and consistent, and produced astonishing breeds of animals and plants after just a few hundred years of selective breeding, much less time than the millions of years generally associated with evolutionary changes in the wild. In the wild, however, selection is seldom this strong and consistent.

Natural selection is extremely important for us to understand today because it has caused the evolution of drug-resistant viruses and bacteria. The overuse of these drugs has selected for the resistant strains, and as a result the drugs are becoming less effective at fighting diseases. Even though this is occurring in the artificial situations of homes and hospitals, it is not artificial selection, because medical researchers are not *trying* to produce resistant germs.

We usually think of crop and livestock breeding as something that we humans are imposing on plants and animals. But, as explained in an earlier chapter, the selection works both ways. Crops and livestock are not simply our slaves. Our selection of high-yielding corn varieties has allowed corn to dominate millions of acres of farmland, giving it an evolutionary fitness that it could never have otherwise achieved.[9] One writer even suggests that some groups of animals "chose" domestication, thereby enhancing their own evolutionary success.[10]

EVOLUTION OF DIVERSITY

Natural selection also allows us to understand the origin of new species. Imagine that the whole Northern Hemisphere was relatively warm, which indeed it was forty million years ago, and that an ancestral population of relatively hairless elephants lived in it. Then along came the Ice Ages. The ancestral elephants that stayed in the warm, sunny south remained hairless and evolved into Indian and African elephants. The ancestral elephants that stayed in the cold north evolved into the hairy elephants we call mammoths and mastodons. So long as the elephant populations did not crossbreed with one another while this process was occurring, natural selection thereby produced different pachyderm species.

There are many different ways of being successful. Natural selection has produced different kinds of adaptations. To survey these adaptations would be to recite the entirety of biology. But I will just use an example from some of my favorite organisms: trees. Oaks, cottonwoods, and alder bushes offer three examples of how a tree can survive and reproduce in its set of environmental conditions.

Oaks live in relatively stable environments, in which an individual tree could reasonably expect to live for centuries without being killed. Under these conditions, it "makes sense" (from a scientist's viewpoint) for a tree to invest a lot of its resources into living for a long time: producing strong wood, for example. Producing strong wood is costly, and that is why an oak tree grows slowly. But what's the rush? The tree will probably get a chance to live hundreds of years.

In contrast, cottonwoods live along rivers, creeks, and lakes. In such places, there is a high risk that an individual tree will get killed by floods. A tree that is capable of living several hundred years will probably not get a chance to do so. The successful trees are those that, like James Dean, live fast and die young. Cottonwoods grow up quickly and may produce seeds before the next big flood comes along. Cottonwood trunks produce flimsy wood; there is not much point in producing wood that will last for centuries. Flimsy wood also tends to have big water tubes (xylem), which carry a lot of water up to the leaves, allowing the cottonwood trees to grow rapidly. Any cottonwood tree that happens to not get killed during its first hundred years will probably fall over from old age before very long.

Then there is a third alternative adaptation that is used by alders and willows, which also live down by waterways. They produce numerous small trunks instead of one large trunk (which is what makes them shrubs instead of trees). No single trunk lives very long, but the underground clump may persist for centuries. Along comes a flood that kills the trunks; the next year, or even sooner, new trunks will sprout from the clump.

We see here three equally good adaptations for tree growth. The oak growth pattern is good for the stable environments of upland forests. The cottonwood and alder growth patterns are good for the unstable environments along waterways. And, as far as I can tell, the two very different growth patterns displayed by cottonwoods and alders seem to be equally good adaptations to a riverside environment.

Natural selection, by favoring whatever works, has produced a prodigious diversity of adaptations, as each evolutionary lineage adapts in different ways to the same conditions or to different conditions.

ONGOING HUMAN EVOLUTION

Humans have continued to evolve biologically ever since our evolutionary origin. However, we have primarily evolved in response not to natural environmental changes but to changes *that we have made to* our environments. We have, in effect, caused our own evolution. For example, as noted in an earlier chapter, some groups of humans figured out agriculture, and those groups became largely dependent on starchy crops. Natural selection favored the individuals in those populations that were better at digesting starch. In addition, some groups of people figured out how to herd animals and drink their milk. While young mammals can digest the lactose sugar found in milk, it is rare for adult mammals to have this ability. But adult humans in populations that herd livestock are more likely to have the ability to digest lactose than adults in populations that do not herd animals for milk.[11] These are genetic, evolutionary changes that have occurred in human populations as a result of natural selection.

In most cases, humans have invented some way to solve the problems we encounter. This includes the problem of lactose intolerance. Humans invented cheese, yogurt, kefir, etc. as ways of preserving milk from rotting and making it more digestible by allowing microbes to break down the lactose. If the environment gets colder, we do not evolve longer hair; we invent clothing. In this way our hairless, tropical bodies can live in periodically or continually cold environments. Technology is our most important adaptation.

THE EVOLUTION OF EVERYTHING

And natural selection applies to everything. Natural selection does not merely explain the evolution of organisms, such as cats, but also the evolution of everything else.[12]

Even technology evolves.[13] If someone comes up with a good idea, the idea spreads to other people. Ideas live in the brains of people just as organisms live in natural environments. Within the collective population of ideas, good ideas reproduce more than bad ideas, just as some organisms reproduce more than others in a population. This, too, is a type of natural selection.

Good ideas do not have to be virtuous or true in order to spread by natural selection within the habitat of the human brain. I think all of us today would say that "Marduk is a great god and he wants us to obliterate you!" is a false idea, but it spread through the minds of the Babylonians and inspired them to conquer other tribes and form the world's first empire. It was a successful idea, however repugnant. The conquerors got slaves and agricultural land, which allowed them to feed the brains in which the Marduk idea lived.

Natural selection might also explain the increasing tendency for scientific papers to be wrong, as I discussed in an earlier chapter. In the competitive world of scientific grants and jobs, it doesn't pay to be right as much as it pays to be first.[14] This is a far cry from the discoverer of natural selection, Charles Darwin, who worked on his idea for twenty-two years before publishing *The Origin of Species*. Darwin said what few modern scientists, rushing to publication, and none of his creationist critics, who published half-baked attacks on evolution, can say: "I have not been hasty in coming to a conclusion."[15]

Therefore, ideas evolve by natural selection.[16] A successful idea is one that spreads the most, regardless of its intrinsic quality. This is why trashy novels often outsell the great novels. We would all agree that Dickens's *A Christmas Carol* is a great novel. It continues to live in countless movies and school plays and has evolved into many forms. This book was Dickens's greatest evolutionary success. But the first edition was not the financial success he expected it to be. He self-published *A Christmas Carol*, earning about £230, rather than the £1,000 he was expecting.[17] Yet *A Christmas Carol* brought Dickens immortality in the world of ideas. I do not know if he has living biological descendants, but he lives on in billions of brains. Ideas also spread more successfully if they come in contact with other ideas. Ideas evolve not just in brains but also in societies of interconnected brains.[18]

Natural selection can also occur *within* computers. There is a whole branch of computer science known as "evolutionary computation."[19] This is how it works. Consider a computer designing an engine component. The programmer feeds in a design for the component—perhaps a good design, perhaps a mere blob. The computer generates slight random mutations of this design. These mutations, along with the original, make up a population that lives not in a forest or an ocean but within the memory of the computer. Then the computer selects the mutant form that functions best within the environment of that particular engine. Next the computer generates another set of mutations and again selects the best one. The computer does this over and over and over, something that computers are really good at doing without getting bored. A few teraflops later, you have what appears to be a brilliantly designed engine component—produced merely by the repeated action of random variation and natural selection.

Natural selection can also occur *between* computers. Cyberspace can be as complex of an ecological community of interactions as a forest or an ocean. And a lot of the "species" out there in cyberspace are dangerous. While "spam" is mostly just annoying, there are viruses that can infect computers. A *biological* virus is not an organism but is a segment of information (DNA or RNA), usually protected by a protein coat. Biological viruses get cells to make copies of them and send them out to infect other cells. In a similar fashion, a *computer* virus is not a piece of hardware but a small amount of software that gets computers to make copies of them and send them out to other computers. Just as some biological viruses are relatively innocuous, some computer viruses are nothing more than harmless replicators. But in many cases the computer viruses leave some instructions behind (often called malware) that will harm the computer or steal personal information from the user (and does the same with any computer to which it is sent). Just as predators and prey are locked in an evolutionary spiral—prey have to evolve better protection from predators, and predators have to evolve better detection of prey— so are computer programmers: some designing ever better protection against viruses, and others designing viruses and malware to get around the protection. Some spam and viruses automatically mutate themselves in a manner that allows them to evade spam and virus detection.

And natural selection might explain the entire universe. Now, the idea I am about to outline is unconventional, and some consider it absurd. But you cannot deny its appeal. It is the "fecund universes" hypothesis of physicist Lee Smolin.[20] Why is it, he asks, that the universal constants (gravity, etc.) are "just right" for the production of stars and therefore planets and therefore life? Of all possible values that these constants could take, the odds against all of them being in just the right range make it nearly impossible for our universe to exist. In fact, it would be nearly impossible for a universe capable of producing stars, much less life, to exist. Only massive stars can produce carbon atoms, which are the basis for life. And only massive stars can, upon collapsing, produce black holes. And yet here we are, in a universe with stars, carbon atoms, and black holes. Some people claim that this is evidence for the existence of God: only God could have done this. Smolin has a different explanation.

Suppose that every time a black hole forms, it has a new universe inside it, an idea proposed by Stephen Hawking.[21] Why not? Nobody can ever know what is inside a black hole anyway, so why not? There could be trillions of fecund black holes in the universe right now. That is, our universe is reproducing. Suppose further that each of these baby universes has slightly different constants than the one that produced the black hole from which it sprang. This is the cosmic equivalent of genetic variation. Many of these baby universes will have constants that make star production impossible. In fact, most of the infant universes will collapse back in on themselves, dying at birth within femtoseconds. By chance, a few of the baby universes will have constants within the right range to allow them to live for billions of years and produce lots of stars and lots of new universes. By this process, the universes with the right constants are the ones that reproduce universes similar to themselves. This is why most universes are similar to our own. This is why our universe should not be so surprising—it is a fairly typical one.

You could think of this as natural selection on a cosmic scale: inferior universes perish, perhaps instantly, and superior universes live a long time and reproduce. For what it's worth, I like it. Of course, since we can never know about these other universes, then this idea remains forever untestable.

You see, we live in a universe and perhaps in a multiverse in which natural selection is everywhere operating. That is why I think that Darwin's insight is the most important idea that humans have ever had. We live in a cosmos where natural law works upon randomness to produce order, but this order is not an outcome that could have been predicted or that is the will of any deity.

DO IT YOURSELF!

Think of something you are familiar with—an object, a process, an art form—and explain how it could evolve.

CHAPTER 18

THE REDISCOVERY OF HUMAN NATURE

O f all the Big Questions, the one about what makes humans human is the question that religion, philosophy, and literature most like to claim as exclusively their own. That's why this general area of study is known as "the humanities." But science has a lot to tell us about what it means to be human, far beyond a mere description of human organs and physiological processes.

There are many erroneous theories of human nature. First there is the fundamentalist religious view, expressed in God's announcement of the Flood to Noah when he said that the thoughts of man's heart are "only evil continually."[1] And Jeremiah said, "The heart is deceitful above all things, and it is exceedingly corrupt."[2] Then there is the old Soviet view that humans have what amounts to a programmable nature. A few generations of government-imposed doctrines and people will have "evolved" into such fine comrades that they will no longer need to be ruled by a dictatorship. And then there is the view that human nature is perfectly good, so long as we can keep out the artificial evil influences of society. But human nature is none of these things.

So what is human nature? Is it good, or is it evil? One of the great insights provided by science is that it is both, and there are good reasons for it. Humans have evolved to usually be good to other humans who are *inside* of their group, and to be evil toward humans *outside* of the group. I refer here to what scientists call *altruism*, within species, rather than *mutualism*, which is cooperation between species. You could define altruism as *doing well by doing good*.[3]

Evolutionary scientists have long puzzled over the enigma of why one animal would be nice to another animal—that is, you could say, why animals sometimes love one another. If evolution consists of just ruthless competition, then why would any animal ever be nice to another? This is not, of course, just a problem for evolution. It is also a problem for creationists. If being nice is a losing strategy, then nice animals would have become extinct even in the few millennia that creationists permit for the history of the universe. God would have to infuse new shots of love into the natural world all the time, keeping love on life support.

Within a close-knit human group (in prehistoric times, this was the village) there were all kinds of reasons for people to be good to one another. Being good to one another helped them to pass on their genes to future generations, which is what evolutionary success is all about.

First, many people in the village were closely related to one another. You can pass on your genes either by having your own children, who receive your genes *directly*, or by helping your relatives to take care of their children, who pass on some of your genes *indirectly*. Helping a relative to successfully raise children is not as effective a way of getting your genes into future generations as is having your own, but it is half as effective, or a quarter as effective, or an eighth as effective, depending on how closely related they are to you. This kind of fitness is called *inclusive fitness* because it includes your relatives' kids, not just your own.[4] Writers have always known that blood is thicker than water, but it took until the twentieth century for scientists to figure it out.

Second, in order to have fitness, you need to have not just offspring but surviving, successful offspring. Evolution rewards not so much those animals who become parents as those who become ancestors. And to do this, your offspring need resources, whether it is food, or whether it is social opportunity. And one of the best ways to get resources (in a social species like humans, chimps, gorillas, or dogs) is to cooperate with other animals in your group. The scientific term for this is *reciprocity*— I do something good for another person, and I receive some kind of resources for it in return. And I mean resources, not just a warm, fuzzy feeling. This is *direct* reciprocity.[5] Writers have always known this. Qoheleth, the writer of the biblical book of Ecclesiastes, said almost three thousand years ago, "For if they fall, the one will lift up his fellow: but woe to him *that is* alone when he falleth; for he hath not another to help him up."[6] But again it took scientists until the twentieth century to figure it out.

Third, a person can be generous to helpless people, even though those people will never be able to repay the favor. The generous person will gain a reputation for trustworthiness among whoever observes or hears about her generosity. This reputation is worth more than money in the bank. If you have a good reputation, the doors to all kinds of resources will be open to you. This is *indirect* reciprocity, since the reward comes indirectly from the observers, not directly from the recipients.[7] The key is for the nice guy to be *conspicuously* generous. Direct reciprocity rewards you with a few good friends to whom you can turn for help, as they can to you; indirect reciprocity rewards you with a whole village of people who are willing to help you.[8] Once again, writers have always known this.

One of literature's most famous characters is Ebenezer Scrooge. He had no kids of his own, nor did he help his nephew raise his kids. No direct or inclusive fitness. He was also mean to people with whom he was in direct contact, and he refused to help the poor, thus eschewing both direct and indirect reciprocity. He was an evolutionary failure.

Altruism is something you do. But emotions such as empathy that reinforce altruism conferred evolutionary fitness as well.[9] Empathy— and love—is something you feel. Natural selection, in many instances, has favored both. This is why it feels good to be good.[10]

Some evolutionary scientists have resisted the idea that altruism could be a product of evolution. They have always pointed out the obvious problem that selfish cheaters could get a free ride on the generous altruism of others. But evolution has given animals two mechanisms to prevent this from happening.

First, there is intelligence. Humans are intelligent enough to remember who can be trusted and who cannot. It is not just the ability to remember facts (intellectual intelligence) but also the ability to pick up on subconscious signals (emotional intelligence). Most humans have pretty good built-in bullshit detectors (as Ernest Hemingway called it), which allow us to identify cheaters who pretend to be cooperators. These signals can be picked up most readily by personal contact. When you have looked someone in the eye, your subconscious mind may be able to figure out whether to trust that person or not.

Second, there is the emotion of sweet revenge. There is no rage so fierce as rage against a cheater—when you are altruistic toward someone and then that person stabs you in the back. The resolution of rage against betrayers of our trust—that is pretty much what a lot of literature is all about.

Of course, there were other social dynamics operating within the prehistoric village. For example, dominance hierarchies. You knew who the alpha males were, and that was that. But alpha males, whether in a human village or in a troop of chimps, get their power not so much from force as from altruism. They have to win support by reciprocity within their close group of associates. Altruism within gangs, if you will.

But outside of the village, natural selection favored violence. Warriors from one village could massacre the inhabitants of another village and feel really good about it. The victimized village carried none of the conqueror's genes (that is, until the conquering men inseminated the women) and were not inclined to offer them any reciprocity.

We have inherited both kinds of instincts: the "good" human nature of altruism within the village, reinforced by the "evil" human nature of revenge, and the "evil" human nature of mercilessness to outsiders. Both sets of emotions—love and hatred—are there, waiting to be used. Throughout human history, we have been extending the boundaries of the group so that now it is customary to feel altruism toward everyone in the world.[11]

ARE ALL HUMANS EQUAL?

Science has something to say about this, too. The short answer is yes. Human races have diverged from a common ancestor only within the last two hundred thousand to one hundred thousand years. That is probably not enough time for any significant evolution to have occurred, except for the abovementioned examples of starch and lactose digestion. Also, natural selection has favored inventiveness, and therefore intelligence, wherever in the whole world physically weak humans have migrated.

The idea that some races, and upper-class people within races, were superior "by nature" (later, by genetics) to other races or to lower-class people within races had long been considered nonsense by honest thinkers. And worse than nonsense. The Nazis attempted to breed pure Aryan people. When, many years after World War II, some of these specially bred people met one another, they found out they were no different from anybody else.[12]

Thomas Jefferson was personally acquainted with many members of European royal families. He not only believed that hereditary monarchies were wrong, but he also knew that they were stupid. He believed strongly that democracy was the right path to follow and also that it was necessary to give every citizen a chance to rise to the top of a democratic society, so as to clear out the stupid people who floated like scum in monarchical societies. As Jefferson wrote in an 1810 letter to Governor John Langdon,[13]

Take any race of animals, confine them in idleness and inaction, whether in a sty, a stable, or a state-room, pamper them with high diet, gratify all their sexual appetites, immerse them in sensualities . . . and banish whatever might lead them to think, and in a few generations they become all body and no mind . . . and this too . . . by that very law by which we are in constant practice of changing the characters and propensities of the animals that we raise for our own purposes. Such is the regimen in raising kings, and in this way they have gone on for centuries. . . . Louis the XVI was a fool. . . . The King of Spain was a fool, and of Naples the same. . . . The King of Sardinia was a fool. All these were Bourbons. The Queen of Portugal, a Braganza, was an idiot by nature. And so was the King of Denmark. . . . The King of Prussia, successor to the great Frederick, was a mere hog in body as well as in mind. Gustavus of Sweden, and Joseph of Austria, were really crazy, and George of England, you know, was in a strait waistcoat. . . . These animals had become without mind and powerless; and so will every hereditary monarch be after a few generations. . . . And so endeth the book of Kings, from all of whom the Lord deliver us.

Science has shown that the racist and oppressive views from the past are simply wrong. And science is one of several significant forces that continue to lead toward the breakdown of barriers between races and between nations.

But sometimes scientists, like anyone else, can act without thinking in a manner that could be called racist. Historically, most scientists have been white males. This is changing rapidly, but science continues to be whiter and more male than the world at large. Now what is a tribe, whether a remnant of prehistory or a participant in the modern world, to think when white male scientists descend upon it and want to study it? Many Native American tribes have refused to participate in DNA testing that is intended to reconstruct their past genetic history. They often fear that DNA testing is another white tool to continue the process of displacing them from their native lands. Many pharmaceutical compounds were first discovered in tropical plants, and the tribes that live in tropical forests around the world have frequently shared their medicinal lore with white investigators. White corporations have gotten very rich as a result of this.

Scientists are really trying to address this problem. Pharmaceutical researchers do not use any knowledge they get from native sources without permission and, when possible, compensation. And recently some scientists, instead of using native peoples as mere informants, have begun to welcome them as coinvestigators. One scientist included the Amazonian native who was most helpful to him, Afukaka Kuikuro of the Kuikuro tribe, as a coauthor on a major paper.[14]

THE RELIGION INSTINCT

Many people consider science to be the enemy of religion. As science advances, religion retreats in desperation. One such person was Thomas Jefferson, who wrote, "The priests of the different religious sects . . . dread the advance of science as witches do the approach of day-light; and scowl on the fatal harbinger announcing the subdivision of the duperies on which they live."[15] God, I wish I'd said that. The more you know about science, the less room there is in the universe for God. You see, *nonreligious scientific explanations always work*. There has never been a time when our assumption of the absence of God has ever caused our explanations to fail.

We have the same brains as those prehistoric people around the campfire. Religion lives in our brains. We always rationalize and sometimes reason. And you expect such a brain to ever be objective about religion? Mine certainly cannot. Nor can the brain of any atheist or any religious believer.

Many consider Darwinian evolution to be the ultimate attack on God. But actually, psychology is religion's worst enemy. We can find a *brain* basis for everything that happens in the human *mind*—happiness, fear, anger, lust, even religious experiences, according to scientists such as Andrew Newberg and Eugene D'Aquili.[16] That is, psychology can explain everything about religious experience and even phenomena such as epilepsy that used to be considered demonic. Moreover, religion is probably ineradicable in the human species.[17] The Soviets tried to eradicate religion and failed. There is a memorable picture of this in Nicholas Mosley's novel *Hopeful Monsters*, in which two children in Soviet Russia had a secret cave where they kept Orthodox ikons.[18] Even some atheists, in their secret moments, might feel drawn to religion as most people are drawn to chocolate.

This exploration of religion leaves unanswered the question that is always in our minds: how do we know what is right and what is wrong? Many people adamantly state that science cannot answer this question. Well, neither can literature or religion. Science can come closer to answering this question than can any other way that I know of. Sam Harris went a little too far when he claimed that we could base our morality upon science,[19] but I think he had the right idea. I believe that whatever promotes altruism (which most of us call love) is good, and that which erodes altruism is evil. Science helps us understand a little better where the human capacity for love came from.

For those of you who want to find clear, *external* evidence for God, I wish you luck. Really, I do. I am afraid that there is, in fact, no such evidence. But as you search for evidence of God, watch out where it might lead. I cannot do better than to end with another quote from Thomas Jefferson: "Question with boldness even the existence of a god; because, if there be one, he must more approve of the homage of reason, than that of blindfolded fear."[20]

DO IT YOURSELF!

I am sure you can think of lots of truly unselfish things that people have done. Choose one of them and think about how the behavior pattern could have evolved by selfish natural selection, even if the person himself or herself was not selfish.

THE ROLE OF SCIENCE IN THE WORLD

What role should science and scientists play in the world? Gone is the time when a scientist can spend his or her time simply studying the natural world for the joy of it, ignoring its consequences for everyone else. As explained previously, scientific discoveries have consequences for the way humans view themselves in the cosmos. Scientific discoveries are likewise essential for humans to make decisions about what to do, as individuals and as societies.

And as a world. In previous centuries, the world could just take care of itself. But all human activities are now so interconnected, and have such an impact on the natural world, that we can no longer let the world take care of itself. Human activities have already, according to Bill McKibben, the author of *Eaarth*, changed Earth into a planet quite different from what it has ever been throughout most of its four-and-a-half-billion-year history.[1] If we humans used the knowledge generated by scientific inquiry, we could prevent our perils.

And what we need most is not, as I said in the introduction, the knowledge that has come from science. It may be the scientific way of thinking itself. Two of the few scientists who were members of Congress were Vern Ehlers (a Republican) and Rush Holt (a Democrat).[2] In Congress, they did not so much use science as use scientific *thinking*.

As these chapters will show, science should play an important role in political decisions, but usually it does not, and the best scientists have public service as their inspiration.

CHAPTER 19

THE SCIENTIST IN A POLITICAL WORLD

The process of science and the knowledge that it has generated are essential for the future survival of the human species, a survival that, alas, rests primarily in the hands of political and economic leaders who are largely ignorant of science, sometimes gleefully so. In such circumstances, a scientist can hardly avoid being political by reaching conclusions that politicians do not like.

ANTISCIENTIFIC POLITICIANS

Political and economic leaders pretty much ignore science. They seem to think that they can declare something to be true and then ignore any scientific evidence to the contrary. Sometimes, political leaders even seem to think that they can override the very laws of nature. Here are some examples.

The legislature of North Carolina decreed that ocean levels would rise at a linear, not an accelerating, rate, regardless of what the oceans might actually decide to do.[1] This reminds me of the old story of King Canute, a ruler of England back during the Viking Danelaw, who told the tide not to rise. But Canute did so in order to demonstrate that even though he was king he could *not* control the forces of nature.

Many of our political leaders have no interest in knowing, for example, the truth about global warming. Few, however, are as hostile as retiring representative Lamar Smith, a Republican from Texas. Incredibly enough, he headed the House Science Committee. He made the claim that he had never seen the evidence that rising carbon dioxide levels in the atmosphere were harmful.[2] In response, the staffers who worked for the Democrats on the committee printed off big piles of scientific papers that proved the very thing of which Smith had never seen evidence. As shown in a startling photograph published in the August 9, 2013, issue of *Science* magazine, Smith simply ignored the pile of papers. So, he has still never seen the evidence.

The hostility that many politicians display toward science is not completely due to exuberant stupidity and greed. One other factor is that some politicians may actually want to feel good about themselves. Consider global warming denialism. If, in fact, the United States is the world's number one carbon emitter (per capita), and most of the effects of global warming will be felt by the poor people of the world (droughts and famines in Africa, destructive monsoons in Asia), then it seems that our pursuit of wealth and luxury is going to kill people. We don't want to think of ourselves as killing people in our pursuit of wealth. So, we invent a story in which our carbon emissions are not, in fact, causing global climate change.[3]

But there is a second reason that politicians ignore science: science is not part of the process of making laws or making money. For example, legislators do their work by writing bills and by slipping riders onto bills that others have written. If you're a legislator who doesn't know how to do this, your career will fail. Business leaders have to find their way through the jungle of shareholders and management as much as some scientists have to find their way through real jungles. Even those political and business leaders who have high esteem for science may not have much time for it.

One of the most extreme examples of a government trying to control science occurred when the Stalinist rulers of the Soviet Union tried to suppress the science of genetics. While the Western world embraced the concept that organisms inherit genes from their ancestors, and that the circumstances of life cannot change these genes, the Soviet Union rejected them. The Soviets apparently believed that an organism can become whatever you force it to become. This means, with regard to humans, that there is no genetically based human nature;

instead, humans can become good little Communists if they are raised to be. The agricultural research of Nikolai Vavilov demonstrated that traits—such as the desirable traits of crops—are determined by genes. This view threatened Soviet philosophy. Instead, the Stalinists preferred to believe the pseudoscience promoted by Trofim Lysenko, who claimed that he could breed new kinds of crops by forcing them to grow in new conditions.[4] Vavilov died in prison while Lysenko enjoyed state-supported renown.[5] Eventually, and quietly, the post-Stalinist Soviet leaders moved away from Lysenko's views. It is impossible to estimate how many Russian peasants starved because Russian scientists, deluded by Lysenko, failed to breed the crops that would have kept them alive.

Most examples of antiscience bias in politics involve conservatives, but liberals are not without their biases too. In most cases, these biases have a trivial effect on the progress of science, as when some activists oppose wind turbines because they sometimes kill birds and bats. (They do, but not as much as global climate change, which wind turbines will help to reduce.) Liberals do have a significant effect on the acceptance of genetically modified organisms (GMOs). Bioengineered organisms are not, in fact, any more dangerous than any other kinds of artificially bred organisms; there is no reason to fear them per se. Of course, like any technology, genetic engineering can be abused. But, particularly in Europe, there are left-wing and highly emotional protests against any and all forms of GMOs. Just try talking to a real liberal about GMOs and you'll get the same feeling that you get from talking to a real conservative about global warming. But even here, the effect is minor compared to the tightly engineered anti-environmental campaign funded by coal and oil corporations and the organizations that they support. Extreme conservatives and extreme liberals may be equally wrong, but the conservatives are more powerful.

Several books have been written about how our political leaders ignore, or are even hostile toward, science, and many scientific hands (delicate hands of lab scientists, gnarly hands of field scientists) have been wrung over why this is so. We scientists tend to blame ourselves, of course. We professors could have made our courses more interesting, for example. And the bored students go on to become our political and business leaders. But I do not accept this argument. There are, in fact, some boring scientists and science teachers. However, most of us teach classes that students find interesting while they take them. But then they forget all about them once they finish. These students are not stupid. It is simply that their work, for example in business or politics (except technology-based businesses), has developed a psychological landscape of indifference or hostility toward science.

THE EXAMPLE OF TOBACCO

I predict that someday soon the truth about global warming will be obvious to everybody, just as we can now see how stupid it was to deny the link between smoking and cancer. But denying the link between smoking and cancer was not due to stupidity. The antiscience lobbyists were and are not stupid. Tobacco companies deliberately funded disinformation campaigns about the dangerous health effects of smoking. Let us take a closer look at this example.

The big tobacco corporations make a lot of money from selling their products, mostly cigarettes. The federal government regulates things that are sold, in an attempt to guarantee that the products do not endanger consumers. Government regulation of product safety started over a hundred years ago and has saved countless lives. Meanwhile, industry has been doing fine selling safe products just as profitably as they would have sold unsafe ones. Tobacco corporations, therefore, want consumers to think that their products are safe. Furthermore, there is an assumption that underlies our entire economic system: consumers are free to choose whether to purchase a product or not. An addicted customer cannot freely choose to *not* purchase the thing to which he or she is addicted. The tobacco corporations have failed in both of these ways: their products are not safe, and their products are addictive. Naturally, tobacco corporations wanted to suppress these truths, and for a long time they got away with it.

The first truth the tobacco corporations wanted to suppress was that smoking can kill you. For decades, people knew that smokers were more likely to get lung cancer, emphysema, and a host of other deadly problems. Smoking is correlated with the increased frequency of lung cancer. Lung cancer, even though it is not the most common form of cancer, was and remains the leading cause of cancer *deaths*, and the overwhelming majority of lung cancer victims were or are smokers.

But how can you prove smoking *causes* lung cancer? One piece of evidence is that smoking comes first and cancer second. It can't very well be the other way around. Or can it? You could always argue that unhealthy people choose to smoke (it makes them feel better, perhaps) and these same unhealthy people are at greater risk of developing lung cancer. Another piece of evidence for causation is that smoking is associated with cancer in the exact part of the body that is exposed to the smoke. Smoking is not as strongly correlated with skin cancer, for example. None of these things absolutely prove causation, but they were strong enough evidence that the federal government required cigarette packages to carry warning labels, beginning in the 1960s, and the government banned cigarette commercials on television and radio in the 1970s. This occurred despite the fact that the tobacco industry hired scientists to dispute the evidence, even though these scientists were not experts in medicine.[6] Most notable was Frederick Seitz, a physicist.

Eventually, proof of causation was found. By the 1990s, scientists had discovered which chemical compound in cigarette smoke (benzopyrene) caused cancer and how it did so: it bound to and mutated the p53 tumor suppressor gene, one of the genes that prevent cancer.[7]

The second truth the tobacco corporations wanted to suppress was that tobacco use is addictive. Scientists have known about nicotine addiction for a long time. However, when Congress subpoenaed the executives of the major tobacco companies to testify in 1994, *all* of the executives raised their hands and swore that they did not believe that nicotine was addictive.[8] About that same time, an employee of one of the tobacco companies copied some of the internal research documents and released them to the *Journal of the American Medical Association*.[9] These documents proved that the tobacco companies knew, from their own secret research, that nicotine was addictive. The corporate researchers even referred to cigarettes as NDDs—nicotine delivery devices. Not only did they know they were selling an addictive product, but they also knew addiction was their product. (The Russell Crowe movie *The Insider* tells this story.)

The tobacco corporations not only knew that addiction was their product, but they also knew that most smokers start smoking when they are young. They accordingly focused some very successful advertising campaigns on young consumers. "Joe Camel" was a particularly successful image that made young people think smoking was cool.

In the late 1990s the federal government and state governments sued the tobacco corporations for the healthcare expenses due to lung cancer and other smoking-related diseases. Tobacco corporations had been making the profits from, and taxpayers footing the bill for, smoking-addiction-induced cancer. Despite the efforts of the tobacco corporations to avoid making these payments, the courts decided in 1998 that the corporations had to pay the expenses and in addition had to stop marketing their products to young people.[10]

A similar thing is happening with the corporate attempts to discredit global warming science. One of the early leaders of these attempts was the same man who fought to discredit the link between smoking and cancer: Frederick Seitz.[11]

The truth gets even more indirect and mysterious. Today, nearly everyone has heard that stress causes numerous health problems, including heart attacks. The scientific research behind the stress–heart disease connection is excellent. But the earliest major researcher who studied the physiology of stress—Hans Selye—got most of his funding from tobacco corporations. The reason is actually quite simple. Numerous things cause heart attacks. Stress is one of them. Smoking is another. Others include poor nutrition and genetic factors. All of these factors interact with one another. And what the tobacco companies wanted to claim, although Selye never actually said this himself (as far as I can tell), was that *stress, not smoking, causes heart attacks*. What Selye did not say, the tobacco companies were eager to say. Their advertisements openly proclaimed that you should smoke to relieve stress.[12]

Hans Selye was certainly a famous scientist, author of thousands of papers and thirty-nine books. Was Selye a liar? It does not appear so. But tobacco companies paid for his research and used his results to lie to the American public. Blood is on their hands, and Selye was their apparently willing tool.

You see, some scientists will say whatever you pay them to say. Far fewer scientists will do this than lawyers and politicians, but there are some. Corporate interests, such as the tobacco, coal, and oil industries, have found these few and used them over and over and over.

SCIENTIFICALLY INFORMED STATESMEN

If you haven't already figured it out, one of my personal heroes is Thomas Jefferson. Everyone knows that he was a genius. When John F. Kennedy hosted Nobel laureates at the White House in 1962, he said that there had never been such a concentration of brain power in the White House dining room except "when Thomas Jefferson dined alone."[13] Jefferson's interest in science exceeded that of every other leading figure of his day, but it was not highly unusual; Enlightenment leaders believed they needed to know about everything, not just about politics. Remember that Benjamin Franklin was quite a scientist too.

In the early 1790s, while he was secretary of state, Jefferson evaluated patent applications. Because only three patents were issued in 1790, statesmen-scholars like Jefferson could give them full attention. When Jacob Isaacs applied for a patent for a still and furnace to distill fresh water from salt water (a potentially important process for long sea voyages), Jefferson required the inventor to assemble his apparatus at the secretary of state's office and demonstrate it personally to Jefferson and to two of the young nation's leading technologists. Isaacs was given five chances to demonstrate that his invention worked, and when he did not satisfy Jefferson and the others, the patent was denied. Clearly, no leading statesman has time to do this anymore. It is the attitude that I wish to focus upon. Jefferson said, regarding patents, "Actual experiment must be required before we can cease to doubt whether the inventor is not deceived by some false or imperfect view of the subject." He insisted that the work of the government should be based on the best science available, including research that avoids bias.[14]

As president, Jefferson arranged for the Lewis and Clark Expedition to explore the newly acquired Louisiana Territory (which extended all the way to what is now Oregon). He was keenly interested in the expedition's scientific discoveries; I have heard that he personally examined the pressed plant specimens sent back by Meriwether Lewis. And in retirement, Jefferson continued the agricultural experiments that he had started long before. He tested out new crops to determine whether they might contribute to the economy of Virginia. His invention of a writing apparatus and a swivel chair, for which he is justly famous, was technology, but his interest in agriculture and the results of the Lewis and Clark Expedition was science, even in the modern sense. The term "science" had a broader meaning in Jefferson's day, yet no one can mistake the meaning of his statement when Jefferson said that science was his passion and that politics was his duty.

Can you find a president or congressman who would say something like this today? It would be difficult to find anyone in politics today who believes, as Jefferson did, that we need science in order to critically test things, whether inventions or ideas, or otherwise our biases might lead us astray.

Even without outright denialism, many special interests like to "spin" the interpretation of data to suit their ends.[15] Scientists cannot ignore the abuse of science by politicians or corporations.

DO IT YOURSELF!

What is your opinion for or against GMOs? Try to spell out your biases. Or try this with any political issue.

CHAPTER 20

WHO IS YOUR FAVORITE SCIENTIST AND WHY?

You are going to be surprised by my answer to this question. You are probably expecting for me to say Darwin or Einstein. And they deserve the admiration that millions of people have for them. But my choice is George Washington Carver.

When I recently taught my graduate research methods course, I told the students that George Washington Carver was a model scientist. None of them had ever heard of him. Most of the students were from Asia and Africa, but even the American students had never heard of him. I believe that science teachers should keep the memory of George Washington Carver alive.

Carver was a great scientist for two main reasons. First, his research focused on turning the agricultural produce of poor rural people into value-added products that would increase their income. He did this in two ways: by improving techniques of production, which increased the farmers' chances of self-sufficiency, and by inventing new markets for their products. Second, he persisted in the face of prejudice. He was good enough to work in a major university, but he could not, because he was black. However, he found his calling at Tuskegee Institute, now Tuskegee University, in Alabama, helping poor rural black farmers in the South.

Carver was born as a slave in Missouri about 1864. Along with his mother and sister, he was kidnapped by Confederate soldiers and sold in Arkansas.[1] The others died, and Carver barely survived. He was returned to his master. Carver's recurring respiratory illness after the

Civil War meant that, instead of doing heavy farm labor with the other sharecroppers and freed slaves, he had time to wander the fields and make observations. He became so knowledgeable about plants that his neighbors called him "the Plant Doctor." His love of plants inspired him to also become a painter. His former master's family helped to educate him, and he traveled to Kansas to continue his primary education. He was deeply impressed by a black woman named Mariah Watkins, who told him he should learn all he could and come back to the South to help his people. When he witnessed the murder of a black man by a white gang, he fled to Minneapolis, Kansas, where he finished high school.

In 1891, Carver was the first black student at Iowa State University, where he studied botany. He graduated in 1894. His mentors were so impressed with him that he stayed to earn a master's degree in 1896. His work at the Agricultural Experiment Station earned him national recognition in the study of fungal diseases of crop plants. In 1896, the president of Tuskegee, educator Booker T. Washington, convinced Carver to join the Tuskegee faculty. Carver remained there for forty-seven years, until his death in 1943.

The soil in Alabama had been depleted by cotton farming. Carver developed systems of crop rotation, in which cotton was alternated with other crops, such as sweet potatoes, so that the soil could build back up. Carver also developed many new uses for crops for food and industry. He developed more than three hundred uses for peanuts, including glue, dyes, ink, varnish, and new foods, which included sauces, but (contrary to popular belief) did not include peanut butter. He did similar research for other Southern crops, including sweet potatoes and pecans. He also developed improvements in adhesives, bleach, buttermilk, linoleum, mayonnaise, paper, plastic, chili sauce, instant coffee, shaving cream, shoe polish, axle grease, and synthetic rubber. He received three patents (one for cosmetics and two for paints). Carver did not, however, keep a good laboratory notebook, and exact formulas for his procedures are largely unknown. To teach the farmers how to use their land better and to create new markets, Carver established an agricultural extension system of advice and laboratory research modeled after the system in Iowa.

Carver was not well known in the United States even though former president Theodore Roosevelt praised him (at Booker T. Washington's funeral) in 1915. He was, however, better known in England, where he was elected to the Royal Society of Arts, one of the few Americans to receive that honor at that time. He became famous when he testified with impressive intelligence before a committee of the US Congress about the many uses he had developed for the peanut. Three American presidents met with him. The crown prince of Sweden studied agriculture under him for three weeks, and the Indian leader Mahatma Gandhi also studied with him. Industrialist Henry Ford financed a laboratory for Carver and worked alongside him in the development of soy-based rubber and synthetic automobile fuel.

After Carver's death on January 5, 1943, President Franklin D. Roosevelt dedicated a national monument to him in Missouri. It is one of the few national sites dedicated to the honor and memory of a black American and perhaps the only one to a botanist.[2]

Carver was also a great model for bringing science and the arts together. I already mentioned his flower paintings. But when he was a young college student, he wrote a poem about gourds and how beautiful they were. It was, to modern tastes, a very overdone poem, but it cannot be denied that he really loved plants. Scientists and science teachers need to love the natural world with this kind of intensity.

My belief in the public responsibility of scientists resulted partly from one of the first scientific seminars I ever attended. It was by Steve Gliessman, who studied the chinampa system of floating-garden agricultural production in Latin America.[3] He was a plant ecologist, who studied under a famous mentor, Cornelius "Neil" Muller at the University of California, Santa Barbara. But unlike his mentor, who studied natural processes in the chaparral and dry oak woodlands, Gliessman studied ways to make food production more efficient—and not for the benefit of large agribusiness but for the sustainability of poor people. All scientists whose work I have studied since that time have, in my mind, had to live up to the kind of standard illustrated by Carver and Gliessman. One of my favorite contemporary scientists is Dan Janzen, a wonderfully crazy genius ecologist, not just for his astonishingly creative insights about the science of ecology but also for how he worked tirelessly to help establish a very large network of national parks in the tropical seasonal forests of Costa Rica. These parks will be his most lasting legacy.

DO IT YOURSELF!

Read up on your favorite scientist, past or present, and explain why you like him or her. Hundreds of biographies of scientists have been written.

CHAPTER 21

AMATEURS AND SPECIALISTS

Most scientists today focus on narrow fields of specialty. They have to; there is so much scientific knowledge that nobody can be competent in every field of science, and rarely in more than one. To do research in any field of scientific study, it is necessary to use very specialized techniques that an amateur could not master. A few scientists have allowed their minds to range over many different fields of thought, even outside of science, when they wrote books for the general public. Examples include Stephen Jay Gould, Rene Dubos, and Lewis Thomas. But they focused their research in their own narrow fields.

But it hasn't always been that way. Before the twentieth century, there were few professional scientists. Instead, there were lots of *naturalists* who studied all of nature rather than just a little sliver of it. Today there are still naturalists, such as David George Haskell,[1] but they tend to be science teachers more often than leaders in scientific research. But two hundred years ago, most scientists devoted themselves to what we today call *natural history*, the study of nature from many different aspects. And they were amateurs, in the good sense of the word.

Let us go back to the days before anything was known about evolution. Gilbert White was, until his death in 1793, the unofficial vicar of Selbourne, in jolly old England. A clergyman. Though he was a country parson, White's passion was natural history.[2]

Unlike most preachers and rich people today, eighteenth- and nineteenth-century English country gentlemen, which included parsons, loved to collect things and study nature. Partly this was because nobody else was doing it. There were very few scientists, so if anybody was going to study trees and birds, it would have to be them. There were few scientific journals (such as *Proceedings of the Royal Society of London*, which is still being published), so most natural history "research" consisted of letters written between one natural historian and another. Parsons also took an interest in natural history because they were among the only people who had time for the pursuit of it. Being a country parson was a fairly comfortable life; you get your short list of church duties out of the way, and you can spend the rest of your time collecting rocks and leaves.

White is most famous for writing a book usually called *The Natural History of Selbourne* (first published in 1789) in which, in the form of letters to friends, he recorded many observations about nature, mostly about animals. The book is just crammed with disorganized information. The book proved to be popular and has remained in print ever since. I own an 1869 copy.

The young Charles Darwin had a copy of White's book. Not only in school but at university, Darwin was much more interested in going out into the woods and collecting stuff than he was in his classical studies. He especially loved to collect beetles. When Darwin's medical training did not work out (he could not stand the sight of blood, especially from stubs of arms amputated without anesthesia), he decided to become a country parson.[3] Now, in case this seems strange, remember that this was before Darwin had any thoughts about evolution. In fact, he was still a conventional Christian. He believed God had created everything, and he wanted to study all these creatures. Darwin read White's book and thought, *This is the kind of life I want to lead.* As it turned out, Darwin inherited a lot of money and could spend his time studying natural history, especially evolution, without having to be a parson.

White's book was not just a compilation of observations. He also tested hypotheses and drew conclusions from them. He was skeptical about some old legends. Here is an example. Where do swallows go in the winter? Legend had it (popularized by the Italian scientist Spallanzani) that swallows hibernated at the bottoms of lakes. White didn't buy it. He suspected that they migrated south, perhaps to Africa. I know it seems obvious to us today that birds migrate. But this fact was not known at the time, except for those obvious examples in which large numbers of birds flew together in huge flocks, like geese and the now extinct passenger pigeons. How could anybody know about the less obvious species, such as swallows? Someone would have to see the swallows in North Africa in the winter and communicate this fact to the natural historians up in England. Or else the natural historians would have to go to North Africa themselves, which White, confined by his ecclesiastical duties, could not do. White and his friends, at least one of whom was in North Africa, gathered whatever information they could. They documented that the swallows were in North Africa precisely when they were absent from England. They concluded that the birds migrated. Of course, they were not sure. Who was to say that they were the same swallows? There was no birdbanding back then.

White drew some further conclusions from his study of swallows. He concluded that while most swallows migrated, some individual swallows do not—that is, not all swallows in a population were the same. This population variation, you will recall, is what makes natural selection possible.

Sometimes White recorded disconnected observations that made no sense to him—but he must have thought that someday there would be an explanation. For example, he wrote about fawns raised by cows, and of a friendship bond forming between a horse and a hen. He also mentioned a mother cat that, when her kittens were killed, suckled a rabbit. He passed on stories of a cat suckling baby squirrels and of a hen taking care of ducklings. These phenomena are still a bit inexplicable, but they must have something to do with altruism, regarding which I wrote earlier. Of course, White also made the error of imputing human thoughts, and even morals, to animals; he described earthworms as being "much addicted to venery" because they spent so much time copulating. White also noted that trees that had been defoliated by insects in late summer, and which grew a new crop of leaves, held onto those leaves longer in the autumn. Today we know this is because new leaves produce more auxin (a hormone) than do old leaves.

White's scientific investigations didn't always work. White tried to draw conclusions from what he observed. He noted that cloudy nights were warmer than clear nights (at similar times of year). Today we know that this occurs because the surface of the earth and everything on it radiate heat into outer space on clear nights, but the sky does not return the heat, while the clouds radiate their own heat back to the earth on cloudy nights. White concluded instead that coldness descended from above the clouds but that the clouds blocked it when they were present. White was wrong, though reasonable.

White also thought about the practical applications of what we now call science. It was one thing to study systematic botany—that is, to classify plants—but it was quite another to use our knowledge of botany to, for example, grow grasses on an eroded hillside. He wrote, "To raise a thick turf on a naked field, would be worth volumes of systematic knowledge."

White also speculated about why there was less leprosy in his day than in earlier centuries. He attributed the decline in leprosy to better agriculture, better food (for example, more vegetables), and cleaner clothes (wearing washable linen instead of long-worn wool). He had no concept of germs, but his speculations were not completely wrong—not bad for a pre-Pasteur natural historian.

White, like everyone else at the time, considered all of the things he observed to be the perfect design of Providence. This idea—that the facts of nature demonstrated the wisdom and goodness of God—was proclaimed most famously by another clergyman, William Paley, in his 1802 book *Natural Theology*, which was another favorite book of the young Charles Darwin. And it was an older Charles Darwin who demolished the concept of natural theology. White did not make a big deal about how natural theology demonstrated the merciful beneficence of Providence. When White's book said that the parasitic insects that live in birds' nests are God's wonderful way of getting baby birds to leave the nest (I am not making this up), it was the editor, not White, who inserted these comments (in the 1869 edition that I own). In one case, White even betrayed just a little hint of doubt about natural theology. Turtles, White noted, lived longer than humans. It made no sense to White for God to give such longevity to a turtle that wastes its time "in joyless torpor."

White communicated his ideas with insight and humor. The introduction to his book contained a letter written by his pet turtle, Timothy (which, it turned out, was actually a female): "From your sorrowful reptile, Timothy."[4] White watched Timothy's behavior not just to write notes in his journal but to also try to figure out why Timothy behaved as (s)he did and maybe to understand how Timothy perceived the world.

Another amateur scientist was the much more famous Henry David Thoreau. He kept detailed observations of the plants and animals around Walden Pond. He kept detailed records of the dates on which wildflowers bloomed, and he counted tree rings to find out whether all the trees in a forest started growing simultaneously or whether they reproduced a little bit at a time. He was as observant as White, but he also counted and measured things, in the spirit of a modern scientist.

The compiled lists of observations made by natural historians in centuries past might turn out to be useful for modern scientific research. Gilbert White's book has a table at the end, in which he summarized many years of observations about the first dates on which flowers bloomed and tree buds opened and birds came back. This is a record of the seasonal patterns of biological events.

Today the scientific study of seasonal patterns is called *phenology*. The term comes from the Greek word for the "study of appearances." Scientists study phenology for several reasons. One is to understand seasonal patterns as an organism's adaptation to its environment. To be a successful oak tree, the tree not only needs to be able to grow in its environment but to also open its buds at the right time. If it opens its buds too early, a late winter frost can kill the young leaves and catkins. If it opens them too late, it will have lost valuable time for growth, and other kinds of trees that were not so tardy in their growth will outgrow them. But another, increasingly important reason for scientists to study phenology is that seasonal patterns, over the decades, are a record of global warming. Some phenologists, such as Richard Primack and Abe Miller-Rushing, have demonstrated that tree buds in parts of the United States open a month earlier than they did back in the 1860s.[5] They reached this conclusion partly on the basis of very thorough records kept by Henry David Thoreau. It is one thing to show, from weather records over that same period of time, that the temperatures have gotten warmer; it is quite another to show that this climatic warming has had any effect on the plants and animals.[6] Since phenology is a new science, there are no long-term data sets compiled by phenologists. Instead, they have to refer to the old records kept by amateur natural historians. This may include the data table in White's book, although to my knowledge nobody has analyzed it scientifically.

Today, science has a great need for amateurs, whom we today call citizen-scientists. There are not enough professional scientists to measure everything everywhere in order to adequately test hypotheses. They have to rely on citizen-scientists to keep records of bird migrations and budburst times and many other things. This costs nothing except for the computer software that allows citizen-scientists to submit their data online. No scientific funding agency could afford to hire a staff big enough to do what citizen-scientists—you!—are happy to do for free.[7] Yes, that means you can join in on scientific research, which, absent your help, might simply never get done.

DO IT YOURSELF!

Find a citizen-science project and join in the fun! There are lots of links available online. But also check with your local nature center or botanic garden. The world of science will be grateful that you did.

CHAPTER 22

SCIENCE IS AN ADVENTURE

Science is not just a discipline—a yoke that allows the ox to advance the cart of knowledge, as I said earlier—but it is also an adventure.

In popular imagination, scientists are not adventurers. Scientists have the image of being timid and socially awkward people. But science is an adventure, even if the adventure takes place inside of a laboratory. Even though I have never gone hang gliding or rock climbing (except for that time I fell down a cliff in Big Bend), I am still an adventurer.

When we think of adventure, we usually think of risky journeys to explore unknown lands. Well, that kind of adventure is pretty much gone. People have been everywhere. So many people have been to the top of Mount Everest that there's even a trail of garbage on the way up.[1] The only way people can have adventures anymore is by doing something new in an old place. People will go to some pretty extreme lengths to do this, like going up many miles into the air in a balloon and jumping out, breaking the sound barrier on the way down, as Felix Baumgartner and Alan Eustace have done.[2] It sounds like *Guinness World Records* has become the new Bible. But science is an adventure that (usually) requires no physical risk but can be every bit as satisfying.

Some scientists, in fact, do things that nearly everyone would consider adventurous, in order to search for new truths. Lonnie Thompson, a climatologist, climbed up high in the Andes to study the melting of the Quelccaya glacier and broke the record for human endurance at such a high elevation.[3] Other scientists live at the South Pole. One of them had to give herself a biopsy while the station she was working in was isolated in the austral winter.[4] Other scientists, looking for new species of plants and animals or making measurements that will help us understand global climate change, trek through steaming jungles while leeches drop down on them like rain from the branches overhead. Most real scientific adventures are not quite as exciting as these, just as real archaeologists do not have adventures quite as thrilling as those of Indiana Jones. In most cases, scientists take only limited physical risks in their pursuits. But the feeling of adventure is still there.

HIDING IN PLAIN SIGHT

Science is one of the kinds of adventures in which people do something new in an old place—that is, we discover new things about the places we have always been to and a new understanding of the things we thought we had always known, things that had been hiding in plain sight. It reminds me of what T. S. Eliot wrote: our explorations will lead us back to the place where we started, and we will understand it for the first time.

Only, I suspect that we will never quite reach the end of our exploring. There will always be something new for scientists to discover. John Horgan wrote in *The End of Science* that science has pretty much figured the universe out.[5] He has gotten a lot of criticism for saying this, but let me stick up for him. In the last couple of hundred years, scientists have discovered how big and how old the universe is, and we have gained a pretty good understanding about how it works at the present time, although the first moments of the big bang remain mysterious. During these last few centuries we have come to understand chemistry and biology here on Earth. While we may yet discover life on another planet, we will find—we must find—that it follows the laws of chemistry that we already know. Horgan is right that there will probably never again be the kinds of breakthroughs such as Edwin Hubble and the red shift, which showed us that the universe is expanding, or Einstein and his two theories of relativity, or Darwin and natural selection. There are still a lot of details to fill in, but we have pretty much illuminated the outlines of reality. All I can say is that there are enough details to fill in that we will never run out of adventures.

We look at plants and animals, and they are the same old plants and animals we have always seen, but with science we understand their DNA and can alter it, and discover the evolutionary history recorded in that DNA. We have the same old brains we have always had, but with science we are beginning to understand how they create our experiences of art, love, and reality. The scientific method is a way to get the physical world to reveal its secrets.

NO CANON, NO DOGMA

We scientists do not sit around and read books and just keep repeating what scientists before us have said. There is no permanent corpus or canon or dogma of scientific truth. Natalie Angier's book *The Canon*, although delightfully written, has a most unfortunate and misleading title.[6] And two of the most famous scientists in history created some confusion by the use of the word "dogma." James Watson and Francis Crick figured out not only the structure of DNA but also how it stored and conveyed genetic information.[7] Crick realized that, apparently, every cell on Earth used DNA in the same way: DNA controls RNA, which controls proteins. He called this the Central Dogma. Although he did not mean to imply it was an unquestionable truth, some people misunderstood him. Crick wrote, "The use of the word dogma caused almost more trouble than it was worth." Religions may have what the Bible calls "the faith delivered once and for all to the saints," but science does not. We are always exploring new truths.

SCIENTIFIC ADVENTURES

My own outdoor adventures in the pursuit of science are not exciting enough to make them into a movie, but they are adventures nevertheless. Among other things, I study alder bushes. Some of these alders live in swamps. A few years ago, I went trudging through a swamp to gather cuttings from their branches to take back to my lab. Out in the swamp, I was up to my butt in the sucking slime. Because the water had very little oxygen, the leaf litter and the corpses of mosquitoes and snapping turtles did not completely decompose. Instead they produced a dark brown glue that stained my clothes like a stygian tea. The bacteria released a putrid scent of hydrogen sulfide. Stupidly, I had no idea how deep the muck was, and my left leg slipped in more deeply as I tried to lift my right. There was nothing to grab onto except poison ivy, the thorny branches of greenbrier and rose, and dead sticks. I was near the bridge of a major US highway; only a few yards from where I stood, hundreds of cars and trucks rumbled past. The motorists who looked down on me might have thought I was crazy if they knew that I was studying the small trees that grew in the swamp. If they knew that I had driven a thousand miles to see these small trees, their suspicions would have been confirmed. But if slurching through a swamp is the only way to discover the answer to a scientific question, then that is what you have to do. You cannot just stand on the side of the road and look at the alder bushes and guess how they would grow or what their DNA is like.

Other scientists stay indoors in order to search for new truths. They work in laboratories, sometimes using materials that outdoor scientists have gathered and sent to them. Scientists sometimes imagine themselves shrinking down to the atomic level and looking around. An adventure, for sure!

And then there are the scientists who stay indoors but enable exploration of places that nobody could reasonably go. Some of them design spacecraft that go to Mars and explore the surface (and even a little under the surface) and beam the information back to Earth. These scientists really get into their work, vicariously, almost as if they were riding around on Mars themselves.

Two scientists whom I immensely admire for their brave creativity are Carl Sagan and Lynn Margulis. They looked about as far out and as far in as anyone could: Sagan studied astronomy, and Margulis studied cells. Neither was satisfied to so limit themselves, however; both of them thought about how their work fitted into the larger picture. This eventually led Sagan to investigate how the brain works,[8] and Margulis to think of Earth as a symbiotic union of organisms.[9] Margulis proposed the now universally accepted idea that complex cells began as a merger of smaller cells. Margulis thought about Earth as a single system, and Sagan knew that this would be true of any other living planet that might be discovered someday. Sagan and Margulis were married for a while. Now they are both gone, but they left a changed scientific world behind them.

Before anyone can explore anything, she or he has to realize that there is something out there to be explored. How many medieval scholars, not to mention peasants, thought of the world as Europe under a glass dome of sky? Perhaps surrounded by a monster-filled sea? Frequently the people who are most ignorant are the most certain about what they believe. As Darwin wrote in the *Descent of Man*, "Ignorance more frequently begets confidence than does knowledge: it is those who know little, and not those who know much, who so positively assert that this or that problem will never be solved by science."[10]

Adventure requires bravery. This applies no less to scientists than to, for example, writers. One example comes from the old Soviet Union. Who were its most famous dissidents? In the 1940s, a famous Russian dissident was the geneticist Nikolai Vavilov, mentioned previously. In the 1960s, one of the most famous dissidents was a physicist, Andrei Sakharov, who warned of the danger of the nuclear technology that physicists—Russian and American—had brought to the world. Of course, there were brave dissident writers as well, such as Boris Pasternak (*Doctor Zhivago*) and Aleksandr Solzhenitsyn (*One Day in the Life of Ivan Denisovitch*). Don't forget Solzhenitsyn. But don't forget Vavilov either.

More good news. Science is not merely an adventure. It is a creative adventure.

Not all adventures are creative. Some people do adventurous things that have been done lots of times before and that other people, often marketers, tell them are adventurous. Riding a motorcycle really fast, for example. That's been done millions of times, about half of them in the street in front of my house. It takes somebody like Evel Knievel to make it really creative: jumping over a canyon on a motorcycle.

Scientific thought is a joyful discipline. Just as exercise is a joyful discipline for the body, and just as we can learn to enjoy eating healthy foods rather than unhealthy ones, the scientific way of thinking is fun. Scientists love science with a passion that many people find puzzling, until they try it for themselves and discover that you don't have to be a scientist to love the richness of a scientific view of the world. Once we begin thinking scientifically, we can look at the world and know that we understand what is happening behind the scenes of what we superficially see. We can feel in charge of our lives. We can look cheerfully on a world from which the demons of ignorance have been chased away. We start noticing all of the beautiful, weird, and complex things that we had previously passed by. Scientific thought can lift us out of just thinking about ourselves, our own persistent problems, and our own temporary pleasures. Scientific thought is a labor of love.

We scientists (including amateur scientists) are brave and creative adventurers. We have more fun than anybody ought to be allowed to have. The kind of excitement that Sir Francis Crick felt when figuring out the structure of DNA, before he was Sir anybody, and that he felt when trying to figure out the brain basis of consciousness, after he was Sir, must have been as great as that felt by another Sir Francis—Francis Drake going around the Horn.

A BEAUTIFUL WORLD

LOVING THE UNIVERSE

E. O. Wilson says that certain themes, such as exploring an undiscovered country and the fight of good against evil, keep coming up in science as well as in literature.[1] Another of these themes is love. One could say that all fictional stories are one story—the story of costly love, told and retold in tens of thousands of ways. Protagonists seek love and antagonists destroy it. There is an important role for love in the stories of science. Science *appeals* to reason. It appeals to *reason*. The human mind *loves* the very act of seeking to understand the world by means of reason. A scientific explanation provides a visceral sense of satisfaction.[2]

Another way in which science is about love is that science explains processes that are important in *human* relationships. How can we say that we love other people if we pollute their air and water, or use technology to oppress them? The natural and social environments are the *media through which we relate to other people*. Science helps us understand these environments and how they work. Science, for example, explains how persistent pollutants such as mercury from coal-fired power plants get concentrated up the food chain, making certain foods like fish toxic. At that point it becomes our responsibility, as lovers, to act upon this understanding, and to stop polluting the planet that belongs to our neighbors, brothers, and sisters. Science also helps us analyze and understand the social environment in which human relationships occur. It may seem a cold mathematical process to calculate a Gini coefficient, which mea-

227

sures inequality. Once we see these Gini coefficients, however—once we are confronted with the fact that the inequality of income in many countries is too high and is increasing—it becomes our responsibility, as lovers, to act upon this understanding and to help our disadvantaged neighbors.

In these ways, science and love are inseparable. This is not the image that most people have of scientists. The caricature of a scientist is a man (sometimes a woman) who is uncomfortable with love and ordinary life, as in the television hit *The Big Bang Theory*. I played the role of such a scientist in a junior high play many years ago. But every scientist I know loves humankind in general and lots of humans in particular and wants to use science as a way of helping other people.

Because science is stories and is inseparable from love, everyone should be able to understand it, since everyone understands stories, love, and love stories. But many or even most people do not see science in this way. You frequently hear the phrase "the hallowed halls of science," which implies that science can only be understood by the initiated who have devoted their lives to understanding its arcane details. I think I can summarize my life's work, and the work of many other scientists and science educators I know, as throwing open the windows of these hallowed halls and inviting people to enter, or at least to look in, to let the winds of public curiosity blow away the incense of scientific mystique. Understandably, people who have not devoted themselves to science can feel uneasy with these hallowed halls. Demagogues parasitize this unease by openly proclaiming that scientists have a secret conspiracy by which we have invented global warming and evolution. They know that people, in general, are uneasy with the idea of "trust us, we're scientists." Scientists and educators like me—through books, blogs, field trips, and YouTube channels[3]—throw open the windows and say, "See the evidence for yourselves." Scientists, writers, and educators like me say that if you can follow the plot of a story, you can understand science.

And this, I believe, is what is most important about science: it gives us stories by which we can understand ourselves and the cosmos. Humans have such an intense hunger for stories that we will believe them even after they have been proven false. Certainly, this is the case with many persistent mythologies. Scientists had better become good storytellers so that the true stories of science can hold their own, at least a little, against the false stories that, in some cases, endanger the survival of the planet. Right now, scientists need to tell the true story of global warming because the false story that global warming is no big deal has spread widely among receptive human minds. For our survival, we humans need not just truth and not just stories but also true stories.

NATURAL LAWS AND MYSTERY

Some people fear that reducing everything to be "merely the product of natural laws" destroys its beauty and wonder. This is not true. Natural laws help us understand the world so that we can see its beauty and wonder.

Some natural laws are things that simply must be so, in any possible universe. I do not think they are just a product of our brains, although, stuck inside of my brain as I am, how would I know? One such law is $1 + 1 = 2$. See what I mean? To question such things is absurd, or at least funny. In the Gilbert and Sullivan comic opera *The Mikado*, first performed in 1885, Nanki-poo was going to be beheaded after a month of wedded bliss. He says, "Well, what's a month? Bah! These divisions of time are purely arbitrary. Who says twenty-four hours make a day? We'll call each second a minute—each minute an hour—each hour a day—and each day a year. At that rate we've about thirty years of married happiness before us!" To which one of the maidens replied, "And, at that rate, this interview has already lasted four hours and three quarters!"

Then there is a second kind of natural laws—laws that operate throughout our universe, but it is possible that there are other universes where the laws are a little different. Why should the strength of gravity be what it is? The charge of an electron? Things like that. They are the "constants." Whatever happened in the first femtoseconds after the big bang presumably determined these constants. Because of them, atoms and stars are the way they are. A carbon atom is a carbon atom is a carbon atom, whether here or thirteen billion light years away from here. There are different isotopes of carbon atoms, but a galaxy far, far away would have the same isotopes. But other universes, should they exist, might not have carbon atoms, or atoms at all.

And then there are the laws that are the result of historical contingency. Contingency is not quite the same as accident. *Contingency* means that once something has happened, future possibilities are constrained. Interestingly, one example is similar to the Mikado joke I quoted above. Why should there be twenty-four hours in a day, and approximately 365 days in a year? Our planet happens to spin at a certain rate and circle the sun at a certain rate. Incidentally, that spin rate is slowing down. Back in dinosaur days, there were over four hundred days a year. At any given time, the length of the day and year are contingent laws.

But even within the world of contingencies, there are universal patterns. They exist because of the unchanging laws of our universe *acting upon* the contingencies of history and evolution.

Consider, for example, advertisement. The reproductive fitness of an organism is increased if its offspring have genetic variation, and to get this variation, the organism has to crossbreed with another organism that is in the same species but not genetically similar to it—that is, crossbreeding is a sweet-spot intermediate between inbreeding and interspecific hybridization. (Yes, I am talking about sex. Now you get to see how a scientist can make sex boring.)

Scientists are not entirely sure why sexual reproduction provides such a fitness boost, but it seems that nearly all types of organisms are sexual. Even bacteria have sex. (You don't want to know.) Now, how does an organism find a mate? It advertises. Humans do this by the way we act and the way we dress. Birds do it too: they have bright feathers and sing complex and beautiful songs. And flowers do it. The whole point of a pretty flower is that its color and scent attract pollinators. Since plants can't copulate, the only remaining option is for something to carry the pollen from the male part of one flower to the female part of another. If this something is an animal, such as a bird or a bee, the flower has to be attractive, not to other flowers but to its pollinators. You could say that "it pays to advertise" is a law of nature in our universe, perhaps in all possible universes. Historical contingency assures that it takes many, many different forms.

While in theory everything that happens may be predictable—if we know everything about energy and matter in a certain system, we can in theory predict what it will do—there is utterly no chance that we will ever do this. The world will always, to a certain extent, appear mysterious to us. In the novel *Jane Eyre*, if we knew what was happening in the trillions of connections in Rochester's brain, we could predict how he would respond to the title character. But anything short of complete knowledge about the states of Rochester's synapses leaves us unable to predict his actions.

Scientists have likened reality to a landscape of hills and vales. You drop a ball on this landscape and try to predict where it will go. It may go to the lowest point, but it may not. It might end up just in the lowest *nearby* point, its further progress blocked by a slight rise in the ground that it encounters, or a stick or a leaf. Now the point is that all the natural laws are acting upon that ball. Gravity, of course. But also the fact that it will roll more rapidly down a steeper hill than a gentler one. It may roll faster if it is a smooth ball than if it is a rough one. Knowing those natural laws might tell you exactly where the ball will go, but in reality the outcome is a surprise. The writer may actually not know how the story is going to turn out when she or he begins writing.

So, you see, we are all bound (all organisms, and all characters in fiction as well) by natural laws, yet there is an astonishing variety of ways in which we fit into those natural laws. There are only a limited number of story lines in the real or the fictional world, and the interest lies in the cleverness of the way in which the biological or fictional characters fit into those story lines. The study of science does not remove beauty from the world any more than the study of harmony and counterpoint removes it from music or the study of perspective removes it from art.

DANGER AND BEAUTY

As I noted previously, science can also be unsettling to people, especially when it threatens their religious beliefs. But it doesn't have to shake up their epistemological underpinnings in order to frighten them. It can simply reveal inconvenient threats. I learned this very early. During elementary summer school, I wrote little articles for the school newspaper back in the days of fragrant purple mimeographs. I had been listening to radio reports about how the abundant rains that year would promote large populations of the mosquito *Culex tarsalis* (probably the first scientific name I learned), which had, in the past, spread viral encephalitis that left some victims, in the words of the reporter, "living vegetables." So I wrote about this. The teacher in charge expurgated almost all of what I'd written and instead wrote, "There will be mosquitoes that will bite you and make you itch, so don't forget the spray!" Mrs. Webb really meant to keep the kids from worrying about being brain-dead after a picnic. But I did learn that a scientist does not need to say everything that could be said. We have to accept the risk of discovering danger in the world if we are to see its beauty.

The scientific process of discovery is, itself, beautiful. One of the best statements of this process is the slogan that was used by the Cherokee Nation for one of its recent holiday events: "Learn from all that I observe."[4] That is a beautiful, and scientific, way to live, even when some of the things that we observe are unsettling.

Another beautiful aspect of science is what E. O. Wilson has called *consilience*.[5] Wilson used the term in the same way that British scholar William Whewell (pronounced *hule*) did almost two hundred years ago. Consilience is a truth that you discern from the convergence of multiple viewpoints. One example of consilience is the idea that an asteroid hit Earth sixty-five million years ago. Numerous, independent lines of evidence converged on this conclusion: the researchers found a sudden decrease in size and diversity of marine microorganisms sixty-five million years ago; they found direct geological evidence of a strong impact, such as certain mineral fragments that can only be produced by strong impacts; and, coup de grace, they found the remnants of the crater itself.

Natural selection is an example of a truth that can be discerned by consilience (even though Whewell himself vigorously rejected Darwinism). You can see how natural selection works if you look at what happens in genetically diverse populations of organisms, *or* in etymologically diverse populations of words and phrases, *or* in psychologically diverse populations of ideas. Evolution is also a concept you can arrive at by consilience. You can conclude that life has evolved over billions of years on this planet by looking at fossils, *or* by looking at DNA, *or* by looking at the process of natural selection that is occurring today. Three independent lines of evidence converge to reveal the truth of evolution. Consilience is beautiful.

In the famous song "Inchworm," Danny Kaye sang that the inchworm spent its time crawling on marigolds without ever stopping to appreciate their beauty. We scientists, in the midst of our measuring, stop and notice the beauty of the world not just once in a while but all the time. Science allows us to appreciate new dimensions of beauty. Einstein said, "The most beautiful thing we can experience is the mysterious. It is the source of all true art and science. He to whom this emotion is a stranger, who can no longer wonder and stand rapt in awe, is as good as dead."[6] A scientist, amateur or professional, hiking in the mountains is, I believe, more likely to notice—and understand—their beauty than someone who is there for the sole purpose of skiing.

There is beauty in the scientific view of the world, both the part we understand and the vastness that we do not. As Lewis Thomas wrote, "We are learning from science how little we know, how still less we understand, and how much there is to learn." It is a beauty that feeds some of our deepest desires. As Thomas also wrote, "Human beings are all right for as long as they are ignorant of ignorance; this is our normal condition. But when we know that we do not know something, we can't stand it."[7]

And once you have had your view of the world changed by science, you cannot go back. Once you know that natural selection is operating everywhere, once you know that leaves are factories that use solar energy, once you start noticing fossils in rocks, once you start recognizing that there are lots of species of trees in a forest, you cannot *unsee* these things.

Science is not just essential for the survival of our species, but it is also one of the best ways that our species has to meet our deepest need: to learn and tell stories about what the world is, who we are, and how we fit into that world.

NOTES

INTRODUCTION: WE NEED SCIENCE, AND WE NEED IT NOW

1. Hugo Mercier and Dan Sperber, *The Enigma of Reason* (Cambridge, MA: Harvard University Press, 2017).

2. Ashley Feinberg, "An 83,000-Processor Supercomputer Can Only Match 1% of Your Brain," Gizmodo, August 6, 2013, https://gizmodo.com/an-83-000-processor-supercomputer-only-matched-one-perc-1045026757 (accessed October 4, 2017).

3. Herman Melville, *Moby Dick* (London: Richard Bentley, 1851), chap. 74.

4. Robert Trivers, *The Folly of Fools: The Logic of Deceit and Self-Deception in Human Life* (New York: Basic Books, 2011).

5. Dave Levitan, *Not a Scientist: How Politicians Mistake, Misrepresent, and Utterly Mangle Science* (New York: Norton, 2017).

6. Karl R. Popper, *The Logic of Scientific Discovery* (New York: Routledge, 1959).

7. National Research Council, *Every Child a Scientist: Achieving Scientific Literacy for All* (Washington, DC: National Academies Press, 1998).

8. Carl Sagan, *The Demon-Haunted World* (New York: Basic Books, 1996).

CHAPTER 1. SCIENCE AND HOW TO RECOGNIZE IT

1. David Wootton, *The Invention of Science: A New History of the Scientific Revolution* (New York: Harper Perennial, 2016).

2. Jeffrey Robins, ed., *The Pleasure of Finding Things Out: The Best Short Works of Richard P. Feynman* (New York: Basic Books, 2005); Richard Dawkins, *The Magic of Reality: How We Know What's Really True* (New York: Free Press, 2011).

3. Karl R. Popper, *The Logic of Scientific Discovery* (New York: Routledge, 1959).

4. Kailash C. Sahu, Jay Anderson, Stefano Casertano, et al., "Relativistic Deflection of Background Starlight Measures the Mass of a Nearby White Dwarf Star," *Science* 356 (2017): 1,046–50.

5. *Climate Change: Information on Potential Economic Effects Could Help Guide Federal Efforts to Reduce Fiscal Exposure* (Washington, DC: US Government Accountability Office, September, 2017), https://www.gao.gov/assets/690/687466.pdf (accessed November 14, 2017).

6. Moody's Investor's Service. "Moody's: Climate Change Is Forecast to Heighten US Exposure to Economic Loss Placing Short- and Long-Term Credit Pressure on US States and Local Governments," press release, November 28, 2017, https://www.moodys.com/research/Moodys-Climate-change-is-forecast-to-heighten-US-exposure-to--PR_376056 (accessed December 4, 2017).

7. D. J. Wuebbles, D. W. Fahey, K. A. Hibbard, et al., eds., *Climate Science Special Report: Fourth National Climate Assessment (NCA4)*, vol. 1 (Washington, DC: US Global Change Research Program, 2017), https://science2017.globalchange.gov/ (accessed November 6, 2017).

8. Mark J. Plotkin, *Tales of a Shaman's Apprentice: An Ethnobotanist Searches for New Medicines in the Amazon Rain Forest* (New York: Viking, 1993).

9. Francis Bacon, *The New Organon* (1620; Cambridge, UK: Cambridge University Press, 2000).

10. Stanley A. Rice, "Roots as Foragers," PlantEd Digital Library, 2012, https://lifediscoveryed.org/r499/roots_as_foragers (accessed October 4, 2017).

11. Stanley A. Rice and Sonya L. Ross, "Smoke-Induced Germination in *Phacelia strictiflora*," *Oklahoma Native Plant Record* 13 (2013): 48–54.

12. France 2France Télévisions, "Pesticides: Des Scientifiques à la Solde de Monsanto?" Francetvinfo.fr, April 10, 2017, http://www.francetvinfo.fr/economie/emploi/metiers/agriculture/pesticides-des-scientifiques-a-la-solde-de-monsanto_2403116.html (accessed October 5, 2017).

13. Stuart Firestein, *Failure: Why Science Is So Successful* (Oxford, UK: Oxford University Press, 2016).

CHAPTER 2. SCIENCE AND FICTION: ORGANIZED COMMON SENSE AND ORGANIZED CREATIVITY

1. C. P. Snow, *The Two Cultures* (Cambridge, UK: Cambridge University Press, 2012).

2. George Gaylord Simpson, *The Dechronization of Sam Magruder* (New York: St. Martin's, 1996); Edward O. Wilson, *Anthill* (New York: Norton, 2011).

3. John Updike, "Visions of Mars," *National Geographic*, January 2008.

4. Alan Lightman, *Ghost* (New York: Vintage, 2008); Alan Lightman, *Mr g: A Novel about the Creation* (New York: Vintage, 2012).

5. Stephen King, *Misery* (New York: Viking, 1987).

6. David Freeman, "Will Mars Colonists Evolve into This New Kind of Human?" NBC News, February 28, 2017, https://www.nbcnews.com/storyline/the-big-questions/mars-colonists-might-evolve-entirely-new-type-human-n708636 (accessed October 25, 2017).

7. *Perry Mason*, season 9, episode 3, "The Case of the Candy Queen," directed by Jesse Hibbs, written by Orville H. Hampton, aired September 26, 1965, on CBS.

8. *Perry Mason*, season 2, episode 15, "The Case of the Foot-Loose Doll," directed by William D. Russell, written by Jonathan Latimer, aired January 24, 1959, on CBS.

9. Ursula W. Goodenough, *The Sacred Depths of Nature* (Oxford, UK: Oxford University Press, 1998); Stanley A. Rice, *Life of Earth: Portrait of a Beautiful, Middle-Aged, Stressed-Out World* (Amherst, NY: Prometheus Books, 2012).

CHAPTER 3. EXPERIMENTING WITH A MOUNTAIN

1. Stanley A. Rice and Fakhri A. Bazzaz, "Quantification of Plasticity of Plant Traits in Response to Light Intensity: Comparing Phenotypes at a Common Weight," *Oecologia* 78 (1989): 502–507.

2. Dror Hawlena, Michael S. Strickland, Mark A. Bradford, et al. "Fear of Predation Slows Plant-Litter Decomposition," *Science* 336 (2012): 1,434–38.

3. Danny Kessler, Klaus Gase, and Ian T. Baldwin, "Field Experiments with Transformed Plants Reveal the Sense of Floral Scents," *Science* 321 (2008): 1,200–202.

4. Matthias Wittlinger, Rüdiger Wehner, and Harald Wolf, "The Ant Odometer: Stepping on Stilts and Stumps," *Science* 312 (2006): 1,965–67.

5. Nina Hahn et al., "Monogenic Heritable Autism Gene *Neuroligin* Impacts *Drosophila* Social Behaviour," *Behavioural Brain Research* 252 (2013): 450–57.

6. G. Shohat-Ophir, K. R. Kaun, R. Azanchi, et al., "Sexual Deprivation Increases Ethanol Intake in *Drosophila*," *Science* 335 (2012): 1,351–55.

7. Tomas Roslin, Bess Hardwick, Vojtech Novotny, et al., "Higher

Predation Risk for Insect Prey at Low Latitudes and Elevations," *Science* 356 (2017): 742–44.

8. Luis W. Alvarez, Walter Alvarez, Frank Asaro, et al., "Extraterrestrial Cause for the Cretaceous-Tertiary Extinction," *Science* 208 (1980): 1,095–107.

9. Frank H. Bormann, Gene E. Likens, D. W. Fisher, et al., "Nutrient Loss Accelerated by Clear-Cutting of a Forest Ecosystem," *Science* 159 (1968): 882–84.

10. E. A. Ainsworth and S. P. Long, "What Have We Learned from 15 Years of Free-Air CO_2 Enrichment (FACE)? A Meta-Analytic Review of the Responses of Photosynthesis, Canopy Properties, and Plant Production to Rising CO_2," *New Phytologist* 165 (2005): 351–71.

11. S. P. Long, E. A. Ainsworth, A. D. Leakey, et al., "Food for Thought: Lower-Than-Expected Crop Yield Stimulation with Rising CO_2 Concentrations," *Science* 312 (2006): 1,918–21.

12. Steven H. Schneider, *Laboratory Earth: The Planetary Gamble We Can't Afford to Lose* (New York: Basic Books, 1998).

CHAPTER 4. WRIGHT AND RONG

1. Stuart Firestein, *Failure: Why Science Is So Successful* (Oxford, UK: Oxford University Press, 2016).

2. Steven Novella, "0.05 or 0.005? P-Value Wars Continue," Science -Based Medicine, August 2, 2017, https://sciencebasedmedicine.org/0-05-or -0-005-p-value-wars-continue/ (accessed October 23, 2017).

3. Roberta B. Ness, *The Creativity Crisis: Reinventing Science to Unleash Possibility* (New York: Oxford University Press, 2014).

4. Richard Harris, *Rigor Mortis: How Sloppy Science Creates Worthless Cures, Crushes Hope, and Wastes Billions* (New York: Basic Books, 2017).

SECTION II: LEGACY OF AN APE'S BRAIN

1. David Chavalarias and John P. A. Ioannidis, "Science Mapping Analysis Characterizes 235 Biases in Biomedical Research," *Journal of Clinical Epidemiology* 63 (2010): 1,205–15.

CHAPTER 5. A WORLD OF ILLUSION

1. Maude W. Baldwin, Yasuka Toda, Tomoya Nakagita, et al., "Evolution of Sweet Taste Perception in Hummingbirds by Transformation of the Ancestral Umami Receptor," *Science* 345 (2014): 929–33.

2. James H. Wandersee and Elisabeth E. Schussler, "Preventing Plant Blindness," *American Biology Teacher* 61 (1999): 82–86.

3. Hannah Faye Chua, Julie E. Boland, and Richard E. Nisbett, "Cultural Variation in Eye Movements during Scene Perception," *Proceedings of the National Academy of Sciences USA* 102 (2005): 12,629–33.

CHAPTER 6. JUST MEASURE IT!?

1. Yuzhang Li, Yanbin Li, Allan Pei, et al., "Atomic Structure of Sensitive Battery Materials and Interfaces Revealed by Cryo-Electron Microscopy," *Science* 358 (2017): 506–10.

2. Nicholas W. Gillham, "Cousins: Charles Darwin, Francis Galton, and the Birth of Eugenics," *Significance* 6 (2009): 132–35.

3. Larry Gonick and Woollcott Smith, *The Cartoon Guide to Statistics* (New York: HarperPerennial, 1993).

4. Stanley A. Rice, *Environmental Variability and Phenotypic Flexibility in Plants* (PhD thesis, University of Illinois at Urbana-Champaign, Department of Plant Biology, 1987). Updated plain English version available at http://www.stanleyrice.com/articles/stanley_rice_thesis_popular_version.pdf (accessed November 1, 2017).

CHAPTER 7. WE SEE LINES WHILE NATURE THROWS US CURVES

1. Lester R. Brown, *The Twenty-Ninth Day: Accommodating Human Needs and Numbers to the Earth's Resources* (New York: Norton, 1978).

2. Joel E. Cohen, "Population Growth and the Earth's Human Carrying Capacity," *Science* 269 (1995): 341–46.

3. Pete Kasperowicz, "National Debt Hits $21 Trillion," *Washington Examiner*, March 16, 2018, https://www.washingtonexaminer.com/news/national-debt-hits-21-trillion (accessed September 6, 2018).

4. Daniel P. Bebber, Francis H. C. Marriott, Kevin J. Gaston, et al., "Predicting Unknown Species Numbers Using Discovery Curves," *Proceedings of the Royal Society B (Biological Sciences)* 274 (2007): 1,651–58.

5. Douglas H. Erwin and James V. Valentine, *The Cambrian Explosion: The Construction of Animal Biodiversity* (New York: Freeman, 2013).

6. Michael J. Benton, "The Origins of Modern Biodiversity on Land," *Philosophical Transactions of the Royal Society of London B (Biological Sciences)* 365 (2010): 3,667–79.

7. Alan De Queiroz, *The Monkey's Voyage: How Improbable Journeys Shaped the History of Life* (New York: Basic Books, 2014).

8. David S. McKay, Everett K. Gibson Jr., Kathie L. Thomas-Keprta, et al., "Search for Past Life on Mars: Possible Relic Biogenic Activity in Martian Meteorite ALH84001," *Science* 273 (1996): 924–30.

9. Tim Sharp, "What Is the Temperature on Mars?" Space.com, November 29, 2017, https://www.space.com/16907-what-is-the-temperature-of-mars.html (accessed September 6, 2018).

10. Lewis Thomas, *The Fragile Species* (New York: Scribner, 1992).

CHAPTER 8. IT'S NOT ALL BLACK AND WHITE

1. Defense Advanced Research Projects Agency (DARPA), "DARPA Z-Man Program Demonstrates Human Climbing Like Geckos," https://www.darpa.mil/news-events/2014-06-05 (accessed September 6, 2018).

2. Michael P. Carlson, "Cyanide Poisoning," http://extensionpublications.unl.edu/assets/html/g2184/build/g2184.htm (accessed September 6, 2018).

3. Yoon Jung Park, "White, Honorary White, or Non-White: Apartheid-Era Constructions of Chinese," *Afro-Hispanic Review* 27 (2008): 123–38.

4. Bryan Sykes, *DNA USA: A Genetic Portrait of America* (New York: Norton, 2012).

5. Encyclopedia of Life, "Encyclopedia of Life: Global Access to Knowledge about Life on Earth," http://www.eol.org (accessed October 28, 2017).

6. Steve Olson, *Mapping Human History: Genes, Race, and Our Common Origin* (New York: Houghton Mifflin Harcourt, 2003).

7. Kenneth R. Miller, *Finding Darwin's God: A Scientist's Search for Common Ground between God and Evolution* (New York: HarperCollins, 1999).

CHAPTER 9. CAUSE AND EFFECT

1. Bum Jin Park, Yuko Tsunetsugu, Tamami Kasetani, et al., "The Physiological Effects of *Shinrin-yoku* (Taking in the Forest Atmosphere or Forest Bathing): Evidence from Field Experiments in 24 Forests across Japan," *Environmental Health and Preventive Medicine* 15 (2010): 18.

2. Wei-win Cheng, Chien-Tsong Lin, Fang-Hua Chu, et al., "Neuro-pharmacological Activities of Phytoncide Released by *Cryptomeria japonica*," *Journal of Wood Science* 55 (2009): 27–31.

3. Edward O. Wilson, *Biophilia* (Cambridge, MA: Harvard University Press, 1986).

4. Vladica M. Veličović, "What Everyone Should Know about Statistical Correlation," *American Scientist* 103 (2015): 26–29.

5. James Orchard Halliwell-Phillipps, *The Nursery Rhymes of England: Collected Chiefly from Oral Tradition*, 4th edition (London: John Russell Smith, 1846), pp. 175–78.

6. "America's Gun Culture in 10 Charts," BBC News, March 21, 2018, https://www.bbc.com/news/world-us-canada-41488081 (accessed September 6, 2018).

7. George W. Cox, *Bird Migration and Global Change* (Covelo, CA: Island Press, 2010).

8. Kenneth Boulding, *The Meaning of the Twentieth Century: The Great Transition* (New York: Harper Collins, 1964), p. 126.

9. Population Reference Bureau, "Data Sheets," https://www.prb.org/datasheets/ (accessed September 6, 2018).

10. T. Talhelm, X. Zhang, S. Oishi, et al., "Large-Scale Differences within China Explained by Rice versus Wheat Agriculture," *Science* 344 (2014): 603–608.

11. David Rindos, *The Origins of Agriculture: An Evolutionary Perspective* (San Diego: Academic Press, 1984).

12. George H. Perry, Nathaniel J. Dominy, Katrina G. Claw, et al., "Diet and the Evolution of Human Amylase Gene Copy Number Variation," *Nature Genetics* 39 (2007): 1,256–60.

13. Linus Pauling, *Vitamin C and the Common Cold* (New York: Freeman, 1970).

14. Linus Pauling, "The Significance of the Evidence about Ascorbic Acid and the Common Cold," *Proceedings of the National Academy of Sciences USA* 68 (1971): 2,678–81.

15. Harri Hemilä, "Does Vitamin C Alleviate the Symptoms of the Common Cold? A Review of Current Evidence," *Scandinavian Journal of Infectious Diseases* 26 (1994): 1–6.

CHAPTER 10. IS BARTHOLO-MEOW INTELLIGENT?

1. Anthony Trewavas, "Aspects of Plant Intelligence," *Annals of Botany* 92 (2003): 1–20.

2. Daniel Chamovitz, *What a Plant Knows: A Field Guide to the Senses* (New York: Farrar, Straus, and Giroux, 2013).

3. "Comment les arbres communiquent entre eux: découvrez le 'réseau internet' de la forêt," Francetvinfo.fr, October 26, 2017, video, http://www .francetvinfo.fr/replay-magazine/france-2/envoye-special/video-le-reseau -internet-de-la-foret_2438099.html (accessed October 26, 2017).

4. Lixiang Li, Haipeng Peng, Jürgen Kurths, et al., "Chaos-Order Transition in Foraging Behavior of Ants," *Proceedings of the National Academy of Sciences USA* 111 (2014): 8,392–97.

5. Bert Hölldobler and Edward O. Wilson, *The Superorganism: The Beauty, Elegance, and Strangeness of Insect Societies* (New York: Norton, 2008).

6. "Birds Attacking Mirrors," MassAudubon.org, https://www .massaudubon.org/learn/nature-wildlife/birds/birds-attacking-windows (accessed September 6, 2018).

7. F. Delfour and K. Marten, "Mirror Image Processing in Three Marine Mammal Species: Killer Whales (Orcinus Orca), False Killer Whales (Pseudorca Crassidens) and California Sea Lions (Zalophus Californianus)," *Behavioural Processes* 53, no. 3 (April 2001): 181–90; Kate Wong, "Dolphin Self-Recognition Mirrors Our Own," *Scientific American*, May 1, 2001, https:// www.scientificamerican.com/article/dolphin-self-recognition/ (accessed September 6, 2018).

8. Bernd Heinrich, *Ravens in Winter* (New York: Vintage, 1989).

9. John M. Marzluff, Jeff Walls, Heather N. Cornell, et al., "Lasting Recognition of Threatening People by Wild American Crows," *Animal Behaviour* 79 (2010): 699–707.

10. Katy Payne, *Silent Thunder: In the Presence of Elephants* (New York: Penguin, 1999).

11. Dale J. Langford, Sara E. Crager, Zarrar Shehzad, et al., "Social Modulation of Pain as Evidence for Empathy in Mice," *Science* 312 (2006): 1,967–70.

12. Emanuela Dalla Costa, Michela Minero, Dirk Lebelt, et al. "Development of the Horse Grimace Scale (HGS) as a Pain Assessment Tool in Horses Undergoing Routine Castration," *PLoS One* 9, no. 3 (2014): e92281, https://doi.org/10.1371/journal.pone.0092281 (accessed September 6, 2018).

13. Isaac Asimov, *Isaac Asimov's Treasury of Humor* (Boston: Houghton Mifflin, 1971), p. 184.

CHAPTER 11. MEASURING WHAT YOU THINK YOU'RE MEASURING

1. Stanley A. Rice and Ian B. Maness, "Brine Shrimp Bioassays: A Useful Technique in Biological Investigations," *American Biology Teacher* 66 (2004): 237–43.

2. Eckard Gauhl, "Photosynthetic Response to Varying Light Intensity in Ecotypes of *Solanum dulcamara* L. from Shaded and Exposed Habitats," *Oecologia* 27 (1976): 278–86.

3. J. M. Clough, James A. Teeri, and Randall S. Alberte, "Photosynthetic Adaptation of *Solanum dulcamara* L. to Sun and Shade Environments. I. A Comparison of Sun and Shade Populations," *Oecologia* 38 (1979): 13–21.

4. Stanley A. Rice and Jennifer R. Griffin, "The Hornworm Assay: Useful in Mathematically-Based Biological Investigations," *American Biology Teacher* 66 (2004): 487–91.

5. Paul Hawken, Amory Lovins, and L. Hunter Lovins, *Natural Capitalism: Creating the Next Industrial Revolution* (New York: Little, Brown, 1999).

6. Eric Davidson, *You Can't Eat GNP: Economics as Though Ecology Mattered* (New York: Basic Books, 2000).

7. "Ratio between CEOs and Average Workers in the World in 2014 by Country," Statista, the Statistics Portal, https://www.statista.com/statistics/424159/pay-gap-between-ceos-and-average-workers-in-world-by-country/ (accessed September 6, 2018).

8. Emily Underwood, "A World of Difference," *Science* 344 (2014) 820–21. See also several articles immediately following this one.

9. Alan Greenspan, "The Challenge of Central Banking in a Democratic Society" (lecture, American Enterprise Institute for Public Policy Research, Washington, DC, December 5, 1996), https://www.federalreserve.gov/boarddocs/speeches/1996/19961205.htm (accessed September 6, 2018).

10. Cristina Silva, "North Korea's Kim Jong Un Is Starving His People to Pay for Nuclear Weapons," *Newsweek*, March 23, 2017, https://www

.newsweek.com/north-koreas-kim-jong-un-starving-his-people-pay-nuclear-weapons-573015 (accessed September 6, 2018).

11. Med Jones, "The American Pursuit of Unhappiness: Gross National Happiness/Well Being (GNH/GNW)," (policy white paper; International Institute of Management, Las Vegas, NV, last updated June 10, 2018), https://www.iim-edu.org/grossnationalhappiness/ (accessed October 17, 2017).

12. Nicholas Enault, "Fusillade à Las Vegas: Trois chiffres ahurissants sur les armes à feu aux États-Unis," Franceinfo, October 3, 2017, http://www.francetvinfo.fr/monde/usa/fusillade-a-las-vegas/infographies-fusillade-a-las-vegas-trois-chiffres-ahurissants-sur-les-armes-a-feu-aux-etats-unis_2401188.html (accessed October 26, 2017).

13. Aamer Madhani, "Baltimore Is the Nation's Most Dangerous City," *USA Today*, February 19, 2018, https://www.usatoday.com/story/news/2018/02/19/homicides-toll-big-u-s-cities-2017/302763002/ (accessed September 6, 2018).

14. Steve Clark, "Census Bureau: Brownsville Poorest City in U.S.," *Brownsville Herald*, November 7, 2013, https://www.brownsvilleherald.com/news/local/census-bureau-brownsville-poorest-city-in-u-s/article_b630f374-475c-11e3-a86e-001a4bcf6878.html (accessed September 6, 2018).

15. "Pollution: Visalia Ranked Second Worst in State," *Visalia Times Delta*, April 24, 2017, https://www.visaliatimesdelta.com/story/news/local/2017/04/24/pollution-visalia-ranked-second-worst-state/100850948/ (accessed September 6, 2018).

16. Wilson Andrews and Alicia Parlapiano, "A History of the C.I.A.'s Secret Interrogation Program," *New York Times*, December 9, 2014, https://www.nytimes.com/interactive/2014/12/09/world/timeline-of-cias-secret-interrogation-program.html. (accessed September 6, 2018).

17. Jeremy Ashkenas, Hannah Fairfield, Josh Keller, et al., "7 Key Points from the CIA Torture Report," *New York Times*, December 9, 2014, https://www.nytimes.com/interactive/2014/12/09/world/cia-torture-report-key-points.html (accessed October 17, 2017).

18. Mazin Sidahmed, "Trump: 'Torture Works,'" *Guardian*, January 26, 2017, https://www.theguardian.com/us-news/2017/jan/26/donald-trump-torture-works (accessed October 17, 2017).

19. Stanley A. Rice, Erica A. Corbett, and Sara N. Henry, "Extent and Variability of Herbivore Damage on Leaves of Post Oaks (*Quercus stellata* Wangenh.) (Fagaceae) in South Central Oklahoma." In review, *Journal of the Botanical Research Institute of Texas*.

20. Michael M. Yartsev, "The Emperor's New Wardrobe: Rebalancing Diversity of Animal Models in Neuroscience Research," *Science* 358 (2017): 466–69.

21. Emily Atkin, "How James Inhofe Snowballed the EPA," *New Republic*, April 23, 2018, https://newrepublic.com/article/148016/james-inhofe -snowballed-epa (accessed September 6, 2018).

22. Josh Levin, "Advantage: Sun," *Slate*, January 16, 2014, http://www .slate.com/articles/sports/sports_nut/2014/01/_2014_australian_open _it_s_110_degrees_at_the_australian_open_why_don_t.html (accessed September 6, 2018).

23. D. J. Wuebbles, D. R. Easterling, K. Hayhoe, et al., "Chapter 1: Our Globally Changing Climate," in *Climate Science Special Report: Fourth National Climate Assessment (NCA4)*, vol. 1 (Washington, DC: US Global Change Research Program, 2017), https://science2017.globalchange.gov/chapter/1/ (accessed November 6, 2017).

24. Xianyao Chen and Ka-Kit Tung, "Varying Planetary Heat Sink Led to Global Warming Slowdown and Acceleration," *Science* 345 (2014): 897–903.

CHAPTER 12. OOPS, I HADN'T THOUGHT OF THAT

1. Robert E. Sorge, Loren J. Martin, Kelsey A. Isbester, et al., "Olfactory Exposure to Males, Including Men, Causes Stress and Related Analgesia in Rodents," *Nature Methods* 11 (2014): 629–32.

2. Richard Dawkins, *River Out of Eden: A Darwinian View of Life* (New York: Basic Books, 1996).

3. Michael J. Behe, *Darwin's Black Box: The Biochemical Challenge to Evolution* (New York: Free Press, 2006).

4. Kenneth R. Miller, "The Mousetrap Analogy or Trapped by Design," http://www.millerandlevine.com/km/evol/DI/Mousetrap.html (accessed September 6, 2018).

5. Charles Darwin, *The Origin of Species by Means of Natural Selection or the Preservation of Favoured Races in the Struggle for Life* (London: Andrew Murray, 1859), chapt. 6.

6. Dawkins, *River Out of Eden*.

CHAPTER 13. EVERYBODY'S BIASED BUT ME

1. "About CRS," Congressional Research Service, last updated April 19, 2018, http://www.loc.gov/crsinfo/about/ (accessed October 28, 2017).

2. Jonathan Weisman, "Non-Partisan Tax Report Withdrawn after GOP Protest," *New York Times*, November 1, 2012, http://www.nytimes.com/2012/11/02/business/questions-raised-on-withdrawal-of-congressional-research-services-report-on-tax-rates.html?_r=0 (accessed October 29, 2017).

3. Slavenka Kam-Hansen et al., "Altered Placebo and Drug Labeling Changes the Outcome of Episodic Migraine Attacks," *Science Translational Medicine* 6 (2014): 218ra5.

4. A. Tinnermann, S. Geuter, C. Sprenger, et al., "Interactions between Brain and Spinal Cord Mediate Value Effects in Nocebo Hyperalgesia," *Science* 358 (2017): 105–108.

5. B. M. Farr and J. M. Gwaltney Jr., "The Problems of Taste in Placebo Matching: An Evaluation of Zinc Gluconate for the Common Cold," *Journal of Chronic Diseases* 40 (1987): 875–79.

6. A. J. Espay et al., "Placebo Effect of Medication Cost in Parkinson Disease: A Randomized, Double-Blind Study," *Neurology* 84 (2015): 794–802.

7. Kelly Servick, "'Nonadherence': A Bitter Pill for Drug Trials," *Science* 346 (2014): 288–89.

8. Kelly Servick, "Outsmarting the Placebo Effect," *Science* 345 (2014): 1,446–47.

9. Stefano Balietti, "Here's How Competition Makes Peer Review More Unfair," The Conversation, August 8, 2016, http://theconversation.com/heres-how-competition-makes-peer-review-more-unfair-62936 (accessed November 7, 2017).

10. Thomas S. Kuhn, *The Structure of Scientific Revolutions* (Chicago: University of Chicago Press, 1962).

11. Luis W. Alvarez et al., "Extraterrestrial Cause for the Cretaceous-Tertiary Extinction," *Science* 208 (1980): 1,095–1,107.

12. "Heavener Runestone," Atlas Obscura, https://www.atlasobscura.com/places/heavener-runestone (accessed September 8, 2018).

13. Wesley Treat, *Weird Oklahoma* (New York: Sterling, 2011).

14. Lynn Sagan, "On the Origin of Mitosing Cells," *Journal of Theoretical Biology* 14 (1967): 255–74.

15. Laasya Samhita and Hans J. Gross, "The 'Clever Hans Phenomenon' Revisted," *Communicative & Integrative Biology* 6, no. 6 (2013): e27122.

16. "Mia Moore: The DOG Who Can Count and Read!" Talent Recap, last updated June 5, 2017, https://www.youtube.com/watch?v=g4T6UaCWk8Y (accessed September 8, 2018).

17. Robert Trivers, *The Folly of Fools: The Logic of Deceit and Self-Deception in Human Life* (New York: Basic Books, 2011).

18. Adrian Desmond and James Moore, *Darwin's Sacred Cause: How a Hatred of Slavery Shaped Darwin's Views on Human Evolution* (New York: Houghton Mifflin Harcourt, 2009).

19. Terry Shropshire, "Bill O'Reilly Says Slaves Who Built White House Were Treated Well," *Atlanta Daily World*, July 27, 2016, https://atlantadailyworld.com/2016/07/27/bill-oreilly-says-slaves-who-built-white-house-were-treated-well/ (accessed October 29, 2017).

20. Herbert Benson, Jeffrey A. Dusek, Jane B. Sherwood, et al., "Study of the Therapeutic Effects of Intercessory Prayer (STEP) in Cardiac Bypass Patients: A Multicenter Randomized Trial of Uncertainty and Certainty of Receiving Intercessory Prayer," *American Heart Journal* 151 (2006): 934–42.

CHAPTER 14. TRUST US, WE'RE SCIENTISTS

1. Naomi Oreskes and Erik M. Conway, *Merchants of Doubt: How a Handful of Scientists Obscured the Truth on Issues from Tobacco Smoke to Global Warming* (New York: Bloomsbury, 2011).

2. Stanley A. Rice, *Green Planet: How Plants Keep the Earth Alive* (New Brunswick, NJ: Rutgers University Press, 2009).

3. A. Baccini, W. Walker, L. Carvalho, et al., "Tropical Forests Are a Net Carbon Source Based on Aboveground Measurements of Gain and Loss," *Science* 358 (2017): 230–34.

4. Dominik Thom, Werner Rammer, and Rupert Seidl, "The Impact of Future Forest Dynamics on Climate: Interactive Effects of Changing Vegetation and Disturbance Regimes," *Ecological Monographs* 87 (2017): 665–84.

5. Seema Jayachandran, Joost de Laat, Eric F. Lambin, et al., "Cash for Carbon: A Randomized Trial of Payments for Ecosystem Services to Reduce Deforestation," *Science* 357 (2017): 267–73.

6. Dennis K. Flaherty, "The Vaccine-Autism Connection: A Public Health Crisis Caused by Unethical Medical Practices and Fraudulent Science," *Annals of Pharmacotherapy* 45 (2011): 1,302–304.

7. E. J. Gangarosa, A. M. Galazka, C. R. Wolfe, et al., "Impact of

Anti-Vaccine Movements on Pertussis Control: The Untold Story," *Lancet* 351 (1998): 356–61.

8. Michael E. Mann, "I'm a Scientist Who Has Gotten Death Threats. I Fear What May Happen under Trump," *Washington Post*, December 16, 2016, https://www.washingtonpost.com/opinions/this-is-what-the-coming -attack-on-climate-science-could-look-like/2016/12/16/e015cc24-bd8c -11e6-94ac-3d324840106c_story.html?utm_term=.5c8914ddcb3b (accessed October 18, 2017).

9. Upton Sinclair, *I, Candidate for Governor: And How I Got Licked* (Berkeley: University of California, 1994), p. 109.

10. Simon Bowers, "Climate-Skeptic US Senator Given Funds by BP Political Action Committee," *Guardian*, March 22, 2015, https://www .theguardian.com/us-news/2015/mar/22/climate-sceptic-us-politician-jim -inhofe-bp-political-action-committee (accessed September 8, 2018).

11. Chris Mooney, "Forget about That Snowball—Here's What Climate Change Could Actually Do to Our Winters," *Washington Post*, March 3, 2015, https://www.washingtonpost.com/news/energy-environment/wp/2015/ 03/03/what-science-and-religion-have-to-say-about-james-inhofes-terrible -snowball-stunt/?utm_term=.247b5e160411 (accessed October 31, 2017).

12. Linda Lear, "The Life and Legacy of Rachel Carson," Rachel Carson.org, 2018, http://www.rachelcarson.org (accessed October 31, 2017).

13. Rachel Carson, *Silent Spring* (New York: Houghton Mifflin, 1962).

14. Paul A. Offit, *Pandora's Lab: Seven Stories of Science Gone Wrong* (Washington, DC: National Geographic, 2017), chap. 6.

15. Rachel Carson, *Silent Spring* (New York: Mariner Books, 2002), pp. 156, 164.

16. Clyde Haberman, "Rachel Carson, DDT, and the Fight against Malaria," *New York Times*, January 22, 2017, https://www.nytimes.com/ 2017/01/22/us/rachel-carson-ddt-malaria-retro-report.html (accessed October 31, 2017).

17. David A. Fahrenthold, "Bill to Honor Rachel Carson on Hold," *Washington Post*, May 23, 2007, http://www.washingtonpost.com/wp-dyn/ content/article/2007/05/22/AR2007052201574.html (accessed October 31, 2017).

18. David Payton, "Rachel Carson (1907–1964)," NASA Earth Observatory, November 13, 2002, https://earthobservatory.nasa.gov/ Features/Carson/Carson3.php (accessed October 31, 2017).

19. Mireya Villareal, "Lawsuit Accuses Monsanto of Manipulating

Research to Hide Roundup Dangers," CBS News, March 15, 2017, https://www.cbsnews.com/news/lawsuit-accuses-monsanto-of-manipulating-research-to-hide-roundup-dangers/ (accessed October 31, 2017).

20. Stuart H. Hurlbert, "Pseudoreplication and the Design of Ecological Field Experiments," *Ecological Monographs* 54 (1984): 187–211.

21. Felisa Wolfe-Simon, Jodi Switzer Blum, Thomas R. Kulp, et al., "A Bacterium That Can Grow by Using Arsenic Instead of Phosphorus," *Science* 332 (2011): 1,163–66.

22. Elizabeth Pennisi, "Concerns about Arsenic-Laden Bacterium Aired," *Science* 332 (2011): 1136–37.

23. David Goodstein, *On Fact and Fraud: Cautionary Tales from the Front Lines of Science* (Princeton, NJ: Princeton University Press, 2010).

24. Oliver Gillie, "Did Sir Cyril Burt Fake His Research on Heritability of Intelligence? Part 1," *Phi Delta Kappan* 58 (1977): 469–71.

25. Mark Israel, *Research Ethics and Integrity for Social Scientists: Beyond Regulatory Compliance* (Thousand Oaks, CA: SAGE, 2014).

26. Woo Suk Hwang, Young June Ryu, Jong Hyuk Park, et al., "Evidence of a Pluripotent Human Embryonic Stem Cell Line Derived from a Cloned Blastocyst," *Science* 303 (2004): 1,669–74.

27. Helen Pearson, "Forensic Software Traces Tweaks to Images," *Nature* 439 (2006): 520–21.

28. J. A. Byrne, D. A. Pedersen, L. L. Clepper, et al., "Producing Primate Embryonic Stem Cells by Somatic Cell Nuclear Transfer," *Nature* 450 (2007): 497–502.

29. Junying Yu, Maxim A. Vodyanik, Kim Smuga-Otto, et al., "Induced Pluripotent Stem Cell Lines Derived from Human Somatic Cells," *Science* 318 (2007): 1,917–20; Kazutoshi Takahashi, Koji Tanabe, Mari Ohnuki, et al., "Induction of Pluripotent Stem Cells from Adult Human Fibroblasts by Defined Factors," *Cell* 131 (2007): 861–72.

30. Haruko Obokata, Teruhiko Wakayama, Yoshiki Sasai, et al., "Stimulus-Triggered Fate Conversion of Somatic Cells into Pluripotency," *Nature* 505 (2014): 641–47.

31. Dennis Normile, "Senior RIKEN Scientist Involved in Stem Cell Scandal Commits Suicide," *Science*, August 5, 2014, http://www.sciencemag.org/news/2014/08/senior-riken-scientist-involved-stem-cell-scandal-commits-suicide (accessed September 7, 2018).

32. Michaela Jarvis, "AAAS Adopts Scientific Freedom and Responsibility Statement," *Science* 358 (2017): 462.

33. Stanley A. Rice, "*Honest Ab: Evolution and Related Topics* (blog)," 2018, http://www.honest-ab.blogspot.com (accessed August 14, 2018).

34. Tom Spears, "Blinded by Scientific Gobbledygook," *Ottawa Citizen*, April 21, 2014, http://ottawacitizen.com/news/local-news/blinded-by -scientific-gobbledygook (accessed November 1, 2017).

35. Andrew M. Sternet al., "Financial Costs and Personal Consequences of Research Conduct Resulting in Retracted Publications," eLife 3 (2014): PMC 4132287.

36. Rick Weiss, "Nip Misinformation in the Bud," *Science* 358 (2017): 427.

CHAPTER 15. TRAPPED

1. Sam Kean, *The Tale of the Dueling Neurosurgeons* (New York: Little, Brown, 2014), chap. 10.

2. Richard Dawkins, *The God Delusion* (New York: Houghton Mifflin, 2006), pp. 127–28.

CHAPTER 16. WHAT EXACTLY DO YOU MEAN?
WHY SCIENTISTS (SHOULD)
CAREFULLY DEFINE THEIR TERMS

1. Ambrose Bierce, *The Unabridged Devil's Dictionary* (Athens, GA: University of Georgia Press, 2002).

2. Benjamin D. Levine and James Stray-Gundersen, ""Living High-Training Low': Effect of Moderate-Altitude Acclimatization with Low-Altitude Training on Performance," *Journal of Applied Physiology* 83 (1997): 102–12.

3. Lawrence P. Greska, Hilde Spielvogel, and Esperanza Caceres, "Total Lung Capacity in Young Highlanders of Aymara Ancestry," *American Journal of Physical Anthropology* 94 (1994): 477–86.

4. Lawrence P. Greska, "Evidence for a Genetic Basis to the Enhanced Total Lung Capacity of Andean Highlanders," *Human Biology* 68 (1996): 119–29; Tatum S. Simonson, Yingzhong Yang, Chad D. Huff, et al., "Genetic Evidence for High-Altitude Adaptation in Tibet," *Science* 329 (2010): 72–75.

5. Stanley A. Rice, "Adaptation," in *Encyclopedia of Evolution* (New York: Facts on File, 2007), p. 1.

6. Samuel H. Taylor, Stephen P. Hulme, Mark Rees, et al.,

"Ecophysiological Traits in C$_3$ and C$_4$ Grasses: A Phylogenetically Controlled Screening Experiment," *New Phytologist* 185 (2009): 780–91.

7. Joseph Felsenstein, "Phylogenies and the Comparative Method," *American Naturalist* 125 (1985): 1–15.

CHAPTER 17. NATURAL SELECTION: THE BIGGEST IDEA EVER

1. Percy Bysshe Shelley, "The Cloud," Poetry Foundation, https://www.poetryfoundation.org/poems/45117/the-cloud-56d2247bf4112 (accessed September 10, 2018).

2. Genesis 1:7 (King James Bible).

3. Leonard Huxley, *Life and Letters of Thomas Henry Huxley, Vol. I* (New York: D. Appleton, 1913), p. 183.

4. Ernst Mayr, *The Growth of Biological Thought: Diversity, Evolution, and Inheritance* (Cambridge, MA: Harvard University Press, 1982).

5. Randall Fuller, *The Book That Changed America: How Darwin's Theory of Evolution Ignited a Nation* (New York: Viking, 2017).

6. Charles Darwin, *The Origin of Species by Means of Natural Selection or the Preservation of Favoured Races in the Struggle for Life* (New York: New American Library, 1958).

7. Immanuel Kant, *Critique of Judgment*, trans. J. H. Bernard (Mineola, NY: Dover, 2005).

8. Darwin, "Introduction," *Origin of Species*.

9. Michael Pollan, *The Botany of Desire: A Plant's-Eye View of the World* (New York: Random House, 2001).

10. Stephen Budiansky, *Covenant of the Wild: Why Animals Chose Domestication* (New York: William Morrow, 1992).

11. Clare Holden and Ruth Mace, "Phylogenetic Analysis of the Evolution of Lactose Digestion in Adults," *Human Biology* 81 (2009): 597–619. Sarah A. Tishkoff, Floyd A. Reed, Alessia Ranciaro, et al., "Convergent Adaptation of Human Lactase Persistence in Africa and Europe," *Nature Genetics* 39 (2006): 31–40.

12. Mark Sumner, *The Evolution of Everything: How Selection Shapes Culture, Commerce, and Nature* (Sausalito, CA: Polipoint, 2010).

13. Richard Dawkins, *The Extended Phenotype: The Long Reach of the Gene*, rev. ed. (Oxford, UK: Oxford University Press, 1999).

14. Paul E. Smaldino and Richard McElreath, "The Natural Selection of Bad Science," *Royal Society Open Science* 3, no. 160384 (September 2016),

http://rsos.royalsocietypublishing.org/content/3/9/160384 (accessed November 7, 2017).

15. Darwin, *Origin of Species*, chap. 1.

16. Susan Blackmore and Richard Dawkins, *The Meme Machine*, rev. ed. (Oxford, UK: Oxford University Press, 2000).

17. Robert Douglas-Fairhurst, "Introduction," in Charles Dickens, *A Christmas Carol and other Christmas Books* (Oxford: Oxford University Press, 2006), pp. vii–xxix.

18. Steven Johnson, *Where Good Ideas Come From: The Natural History of Innovation* (New York: Penguin, 2010).

19. David B. Fogel, *Evolutionary Computation: The Fossil Record* (Hoboken, NJ: John Wiley, 1998). This book is outdated but explains what makes these algorithms evolution.

20. Lee Smolin, *The Life of the Cosmos* (New York: Oxford University Press, 1999).

21. Stephen Hawking, *Black Holes and Baby Universes* (New York: Bantam, 1994).

CHAPTER 18. THE REDISCOVERY OF HUMAN NATURE

1. Genesis 6:5 (King James Bible).

2. Jeremiah 17:9 (World English Bible).

3. Stanley A. Rice, *Life of Earth: Portrait of a Beautiful, Middle-Aged, Stressed-Out World* (Amherst, NY: Prometheus Books, 2012), chap. 6.

4. Lee Alan Dugatkin, "Inclusive Fitness Theory from Darwin to Hamilton," *Genetics* 176 (2007): 1,375–80.

5. Robert L. Trivers, "The Evolution of Reciprocal Altruism," *Quarterly Review of Biology* 46 (1971): 35–57.

6. Ecclesiastes 4:10 (King James Bible).

7. Martin A. Nowak and Karl Sigmund, "The Evolution of Indirect Reciprocity," *Nature* 437 (2005): 1,291–98.

8. Martin A. Nowak and Roger Highfield, *Super Cooperators: Altruism, Evolution, and Why We Need Each Other to Succeed* (New York: Free Press, 2011).

9. Frans De Waal, *The Age of Empathy: Nature's Lessons for a Kinder Society* (New York: Crown, 2010).

10. Dacher Keltner, *Born to Be Good: The Science of a Meaningful Life* (New York: Norton, 2009); Michael Shermer, *The Science of Good and Evil: Why People Cheat, Gossip, Care, Share, and Follow the Golden Rule* (New York: Henry Holt, 2004).

11. Steven Pinker, *The Better Angels of Our Nature: Why Violence Has Declined* (New York: Penguin, 2012).

12. Mark Landler, "Results of Secret Nazi Breeding Program: Ordinary Folks," *New York Times*, November 7, 2006, http://www.nytimes.com/2006/11/07/world/europe/07nazi.html (accessed July 8, 2011).

13. Thomas Jefferson to Gov. John Langdon, March 5, 1810, in *Thomas Jefferson: Writings*, ed. Merrill D. Peterson (New York: Library of America, 1984), pp. 1,218–22.

14. Michael J. Heckenberger, J. Christian Russell, Carlos Fausto, et al., "Pre-Columbian Urbanism, Anthropogenic Landscapes, and the Future of the Amazon," *Science* 321 (2008): 1,214–17.

15. Thomas Jefferson to José Corrêa da Serra, April 11, 1820, in Founders Online, National Archives, https://founders.archives.gov/documents/Jefferson/98-01-02-1213 (accessed September 6, 2018).

16. Andrew Newberg, Eugene D'Aquili, and Vince Rause, *Why God Won't Go Away: Brain Science and the Biology of Belief* (New York: Ballantine, 2002).

17. Jesse Bering, *The Belief Instinct: The Psychology of Souls, Destiny, and the Meaning of Life* (New York: Norton, 2012).

18. Nicholas Mosley, *Hopeful Monsters* (Elmwood, IL: Dalkey Archive Press, 1991).

19. Sam Harris, *The Moral Landscape: How Science Can Determine Human Values* (New York: Free Press, 2011).

20. Thomas Jefferson to Peter Carr, August 10, 1787, in Thomas Jefferson Foundation, http://tjrs.monticello.org/letter/1297 (accessed September 10, 2018).

SECTION IV: THE ROLE OF SCIENCE IN THE WORLD

1. Bill McKibben, *Eaarth: Making a Life on a Tough New Planet* (New York: St. Martin's Griffin, 2011).

2. Emily Langer, "Vern Ehlers, Nuclear Physicist Who Went to Congress, Dies at 83," *Washington Post*, August 17, 2017, https://www.washingtonpost.com/local/obituaries/vern-ehlers-nuclear-physicist-who-went-to-congress-dies-at-83/2017/08/17/ca6010c8-82d1-11e7-b359-15a3617c767b_story.html?utm_term=.248709a34a2e (accessed September 9, 2018); Claudia Dreifus, "A Conversation with: Rush Holt; at Last, a

Politician Who Knows Quantum Mechanics," *New York Times*, November 24, 1998, https://www.nytimes.com/1998/11/24/science/a-conversation-with -rush-holt-at-last-a-politician-who-knows-quantum-mechanics.html (accessed September 9, 2018).

CHAPTER 19. THE SCIENTIST IN A POLITICAL WORLD

1. Leigh Phillips, "US Northeast Coast Is Hotspot for Rising Sea Levels," *Nature*, June 24, 2012, https://www.nature.com/news/us-northeast -coast-is-hotspot-for-rising-sea-levels-1.10880 (accessed November 9, 2017).

2. Kelly Servick, "House Subpoenas Revives Battle over Air Pollution Studies," *Science* 341 (2013): 604.

3. Roland Bénabou and Jean Tirole, "Belief in a Just World and Redistributive Politics," *Quarterly Journal of Economics* 121 (2006): 699–746; Matthew Feinberg and Robb Willer, "Apocalypse Soon? Dire Messages Reduce Belief in Global Warming by Contradicting Just World Beliefs," *Psychological Science* 22 (2011): 34–38.

4. Valery Soyfer, *Lysenko and the Tragedy of Soviet Science* (New Brunswick, NJ: Rutgers University Press, 1994).

5. Peter Pringle, *The Murder of Nikolai Vavilov: The Story of Stalin's Persecution of One of the Great Scientists of the Twentieth Century* (New York: Simon and Schuster, 2008).

6. Naomi Oreskes and Erik M. Conway, *Merchants of Doubt: How a Handful of Scientists Obscured the Truth on Issues from Tobacco Smoke to Global Warming* (New York: Bloomsbury, 2011).

7. Mikhail F. Denissenko, Annie Pao, Moon-shong Tang, et al., "Preferential Formation of Benzo[a]pyrene Adducts at Lung Cancer Mutational Hotspots in P53," *Science* 274 (1996): 430–32.

8. Ibid.

9. J. Slade, L. A. Bero, P. Hanauer, et al., "Nicotine and Addiction: The Brown and Williamson Documents," *Journal of the American Medical Association (JAMA)* 274 (1995): 225–33.

10. Barry Meier, "Cigarette Makers and States Draft a $206 Billion Deal," *New York Times*, November 14, 1998, http://www.nytimes.com/1998/11/14/us/cigarette-makers-and-states-draft-a-206-billion-deal.html (accessed November 9, 2017).

11. Oreskes and Conway, *Merchants of Doubt*.

12. Alix Spiegel, "The Secret History behind the Science of Stress,"

National Public Radio, July 7, 2014, https://www.npr.org/sections/health-shots/2014/07/07/325946892/the-secret-history-behind-the-science-of-stress (accessed November 9, 2017).

13. John F. Kennedy, "Extract from John F. Kennedy's Remarks at Dinner Honoring Nobel Prize Winners of the Western Hemisphere" (speech, White House, Washington, DC, April 29, 1962), Gerhard Peters and John T. Woolley, eds., American Presidency Project, http://tjrs.monticello.org/letter/1856 (accessed September 10, 2018).

14. Carroll W. Pursell, *Technology in America: A History of Individuals and Ideas*, 2nd ed. (Boston: MIT Press, 1990).

15. Ibid.

CHAPTER 20. WHO IS YOUR FAVORITE SCIENTIST AND WHY?

1. Gary R. Kremer, ed., *George Washington Carver: In His Own Words* (Columbia: University of Missouri Press, 1991).

2. Ibid.

3. S. R. Gliessman, R. E. Garcia, and M. A. Amador, "The Ecological Basis for the Application of Traditional Agricultural Technology in the Management of Tropical Agro-Ecosystems," *Agro-Ecosystems* 7 (1981): 173–85.

CHAPTER 21. AMATEURS AND SPECIALISTS

1. David George Haskell, *The Forest Unseen: A Year's Watch in Nature* (New York: Penguin, 2013).

2. Richard Mabey, *Gilbert White: A Biography of the Author of the Natural History of Selborne* (Charlottesville: University of Virginia Press, 2007).

3. Adrian Desmond and James Moore, *Darwin: The Life of a Tormented Evolutionist* (New York: Time Warner, 1992), p.27.

4. Verlyn Klinkenborg, *Timothy: Or, Notes of an Abject Reptile* (New York: Vintage, 2007).

5. Richard B. Primack, *Walden Warming: Climate Change Comes to Thoreau's Woods* (University of Chicago Press, 2014); A. J. Miller-Rushing and R. B. Primack, "Global Warming and Flowering Times in Thoreau's Concord: A Community Perspective," *Ecology* 89 (2008): 332–41.

6. Charles C. Davis, Charles G. Willis, Bryan Connolly, et al., "Herbarium Records Are Reliable Sources of Phenological Change Driven

by Climate and Provide Novel Insights into Species' Phenological Cueing Mechanisms," *American Journal of Botany* 102 (2015): 1,599–609.

7. "Projects," Citizen Science Alliance, https://www.zooniverse.org/projects (accessed November 14, 2017).

CHAPTER 22. SCIENCE IS AN ADVENTURE

1. Athena Yenko, "Mount Everest Becomes Highest Garbage Dump in the World,'" *Tech Times,* June 18, 2018, https://www.techtimes.com/articles/230453/20180618/mount-everest-becomes-highest-garbage-dump-in-the-world.htm (accessed September 9, 2018).

2. John Markoff, "Parachutist's Record Fall: Over 25 Miles in 15 Minutes," *New York Times,* October 24, 2014, https://www.nytimes.com/2014/10/25/science/alan-eustace-jumps-from-stratosphere-breaking-felix-baumgartners-world-record.html (accessed September 9, 2018).

3. Kevin Krajick, "Ice Man: Lonnie Thompson Scales the Peaks for Science," *Science* 298 (2002): 518–22.

4. Patricia Sullivan, "Obituary: Jerri Nielsen; Doctor Battled Cancer at South Pole," *Washington Post,* June 26, 2009, http://www.washingtonpost.com/wp-dyn/content/article/2009/06/24/AR2009062403094.html (accessed September 9, 2018).

5. John Horgan, *The End of Science* (New York: Broadway Books, 1997).

6. Natalie Angier, *The Canon: A Whirligig Tour of the Beautiful Basics of Science* (New York: Mariner Books, 2008).

7. Francis Crick, *What Mad Pursuit: A Personal View of Scientific Discovery* (New York: Basic Books, 1990).

8. Carl Sagan, *The Dragons of Eden: Speculations on the Evolution of Human Intelligence* (New York: Random House, 1977).

9. Lynn Margulis, *Symbiotic Planet: A New Look at Evolution* (New York: Basic Books, 1998).

10. Charles Darwin, *The Descent of Man, and Selection in Relation to Sex* (New York: Penguin, 2004).

EPILOGUE: A BEAUTIFUL WORLD

1. Edward O. Wilson, *Letters to a Young Scientist* (New York: Norton, 2013).

2. Mario Livio, *Why? What Makes Us Curious* (New York: Simon and Schuster, 2017).

3. "StanEvolve: The Darwin Channel," Stanley A. Rice, last updated April 20, 2018, http://www.youtube.com/StanEvolve (accessed August 14, 2018).

4. "Cherokee Nation Principal Chief Presents State of the Nation Address on Saturday," Cherokee Nation, September 2, 2009, http://www.cherokee.org/News/Stories/24056 (accessed September 12, 2018).

5. Edward O. Wilson, *Consilience: The Unity of Knowledge* (New York: Knopf, 1998).

6. Albert Einstein, *Living Philosophies* (New York: Simon and Schuster, 1931).

7. Lewis Thomas, *The Fragile Species* (New York: Scribner, 1992).

INDEX